LABORATORY MANUAL

PRINCIPLES OF

ELECTRONIC COMMUNICATION SYSTEMS

DAVID L. HEISERMAN

GLENCOE

McGraw-Hill

New York, New York   Columbus, Ohio   Woodland Hills, California   Peoria, Illinois

Higher Education

LABORATORY MANUAL PRINCIPLES OF ELECTRONIC COMMUNICATION SYSTEMS
SECOND EDITION

1 2 3 4 5 6 7 8 9 0 QPD/QPD 0 9 8 7 6 5 4 3

ISBN 0–07–294315–7

Publisher: *David T. Culverwell*
Developmental editor: *Patricia Forrest*
Senior marketing manager: *Roxan Kinsey*
Project manager: *Sheila M. Frank*
Senior production supervisor: *Laura Fuller*
Media technology producer: *Janna Martin*
Senior coordinator of freelance design: *Michelle D. Whitaker*
Cover: (l) *Larry Hamill*, (c) *Jeffry Myers/Index Stock*, (r) *Caroline Cronin*; **iii** *Corbis*; **iv-v** *PhotoDisc, Inc.*
Supplement producer: *Brenda A. Ernzen*
Compositor: *The GTS Companies/York, PA Campus*
Typeface: *10/12 Janson*
Printer: *Quebecor World Dubuque, IA*
SweetHaven Publishing Services www.sweethaven.com

Glencoe software support: 1-800-331-5094, option 3

Printed in the United States of America.

www.mhhe.com

COMPUTER REQUIREMENTS AND SETUP INSTRUCTIONS

Minimum Computer Requirements

Windows® 98 or higher
24 MB RAM
10 MB hard drive Storage
800 × 600 display resolution
16-bit color
16x CD-ROM drive

Start-Up Instructions

The CD-ROM uses an autostart feature that automatically starts the appropriate software procedure. If this software has never been run on the computer, it automatically starts the installation process. But if the software was installed at some earlier time, it automatically starts the lab simulation application.

Note: If you have just installed this software for the first time, you must remove and then replace the CD-ROM in order to start the lab simulations.

ABOUT THE AUTHOR

David L. Heiserman is a fifteen-year veteran of the classroom, conducting classes in electronics technology and technical mathematics at the Ohio Institute of Technology, Franklin University (Columbus, Ohio), and Columbus State Technical College.

He studied applied mathematics and psychology at Ohio State University and earned graduate credit in biophysics. His knowledge of electronics stems from his years as aviation radar weapons control specialist in the Navy. He holds two U.S. patents for micro-miniature robotic grippers.

Heiserman has authored over thirty books on technical and scientific subjects, including three on digital electronics and microprocessor instruction sets. He is the founder and education director for Free-Ed.Net, an Internet school that offers career and academic courses at no cost to the public.

CONTENTS

Computer Requirements and
Setup Instructions iii

Preface vii

Project 1 Passive Filters
      (A Prep Project) 1

Project 2 Passive Filters
      (A Hands-On Project). 9

Project 3 Active Filters
      (A Prep Project) 19

Project 4 Active Filters
      (A Hands-On Project). 23

Project 5 Ceramic Filters
      (A Prep Project) 29

Project 6 Ceramic Filters
      (A Hands-On Project). 33

Project 7 Bode plotter
      (A multiSIM Project) 37

Project 8 Sweep Generators
      (An Extended Project). 41

Project 9 Fourier Analysis
      (An Extended Project). 47

Project 10 AM Spectral Analysis
      (An Extended Project). 51

Project 11 AM Modulation Index
      (A multiSIM Project) 55

Project 12 Amplitude Modulation I
      (A Prep Project) 59

Project 13 Amplitude Modulation I
      (A Hands-On Project). 63

Project 14 Amplitude Modulation II
      (A Prep Project) 67

Project 15 Amplitude Modulation II
      (A Hands-On Project). 73

Project 16 Diode AM Detector
      (A Prep Project) 79

Project 17 Diode AM Detector
      (A Hands-On Project). 83

Project 18 AM Diode Detector
      (A multiSIM Project) 89

Project 19 SSB Modulator
      (A Prep Project) 93

Project 20 SSB Demodulator
      (A Hands-On Project). 99

Project 21 Lattice Modulator
      (A multiSIM Project) 103

Project 22 FM Spectrum Analysis
      (An Extended Project). 107

Project 23 Frequency-Shift Keying
      (A multiSIM Project) 111

Project 24 The Bessel Function Block
      (A commSIM Project). 115

Project 25 Frequency Modulation
      (A Prep Project) 119

Project 26 Frequency Modulation
      (A Hands-On Project). 123

Project 27 Analog Frequency Multiplier
      (A multiSIM Project) 127

Project 28 Varactor Modulator
      (An Extended Project). 131

Project 29 Frequency Demodulation
      (A Prep Project) 137

Project 30 Frequency Demodulation
      (A Hands-On Project). 141

Project 31 PLL Operation
      (A Prep Project) 145

Project 32 PLL Operation
      (A Hands-On Project). 149

Project 33 Crystal Oscillator
      (A Prep Project) 153

Project 34 Crystal Oscillator
      (A Hands-On Project). 157

Project 35 Tuned Amplifiers
      (A Prep Project) 161

Project 36   Tuned Amplifiers
(A Hands-On Project) 165

Project 37   Tuning RF Amplifiers Circuits
(An Extended Project) 169

Project 38   Frequency Multipliers
(A Prep Project) 173

Project 39   Frequency Multipliers
(A Hands-On Project) 177

Project 40   Impedance-Matching Networks
(A Prep Project) 181

Project 41   Impedance-Matching Networks
(A Hands-On Project) 185

Project 42   D/A Conversion
(A Prep Project) 189

Project 43   D/A Conversion
(A Hands-On Project) 193

Project 44   A/D Conversion
(A Prep Project) 197

Project 45   A/D Conversion
(A Hands-On Project) 201

Project 46   Pulse-Code Modulation
(An Extended Project) 205

Project 47   Pulse-Width Modulation
(A Prep Project) 209

Project 48   Pulse-Width Modulation
(A Hands-On Project) 213

Project 49   The Compander Block
(A commSIM Project) 217

Project 50   Sample-and-Hold Circuit
(A multiSIM Project) 221

Project 51   Frequency Converter
(A Prep Project) 225

Project 52   Frequency Converter
(A Hands-On Project) 229

Project 53   Simple AM Receiver
(A Hands-On Project) 233

Project 54   Superheterodyne Receivers
(An Extended Project) 237

Project 55   AM/FM Radio Receiver
(An Extended Project) 241

Project 56   Signals and Noise
(A commSIM Project) 245

Project 57   Data Multiplexer
(A Prep Project) 249

Project 58   Data Multiplexing
(A Hands-On Project) 253

Project 59   Data Demultiplexing
(A multiSIM Project) 257

Project 60   Analog Multiplexer
(A multiSIM Project) 261

Project 61   RF Switch
(A commSIM Project) 265

Project 62   Parallel/Serial Conversion
(An Extended Project) 269

Project 63   Modems
(An Extended Project) 273

Project 64   Binary Phase-Shift Keying
(A multiSIM Project) 277

Project 65   Parity Generator/Checker
(A Prep Project) 281

Project 66   Parity Generator/Checker
(A Hands-On Project) 285

Project 67   Pinging Internet Servers
(A Hands-On Project) 289

Project 68   Antenna Voltage and Current
(A Prep Project) 293

Project 69   Antenna Voltage and Current
(A Hands-On Project) 297

Project 70   Antenna Impedance Matching
(An Extended Project) 301

Project 71   Microwave Systems
(A Virtual Project) 305

Project 72   Pulse-Tone Dialer
(An Extended Project) 309

Project 73   Fiber-Optic Communication
(A Hands-On Project) 313

Project 74   Television System
(An Extended Project) 317

Project 75   Television Remote Controls
(A Prep Project) 319

Project 76   Television Remote Controls
(A Hands-On Project) 323

Project 77   Lissajous Patterns
(A multiSIM Project) 327

Appendix A   Composite Parts
and Equipment Lists 331

Appendix B   Semiconductor Pinouts 333

PREFACE

The domain of modern electronic communications is expanding rapidly—into realms of higher frequencies, more and better consumer products, and entirely new digital-based products and services. And just as the technology is becoming more sophisticated and wider ranging, so must the methods for preparing the engineers and technicians who make it all work. This laboratory manual, and its accompanying simulation software and Internet support, present a significant step in that direction.

In the mid-1990s, the most sophisticated and expensive radio-frequency equipment could operate only in the hundreds-of-megahertz range. Today, however, there are consumer wireless devices operating in the thousands-of-megahertz range. Not long ago, the telephone was a relatively simple instrument that carried low-frequency, narrow-channel voice signals. Now the same device and its supporting systems are expected to handle voice, fax, and computer telecommunications. Millions of personal computers in homes, schools, and businesses are interconnected on local networks and over the entire world via the Internet. Communications between computers is no longer an option; rather, it is a vital part of the development of information technology.

So it is clear that communications technology is indeed growing and changing very rapidly; and it is equally evident that the methods and resources for teaching communications technology must keep pace with these changes. This laboratory system (comprised of the hardcopy manual and software projects) answers this demand by offering a far winder range and number of projects than possible with the traditional approach. The sequence of projects follows the textbook, chapter-by-chapter, supporting the theory and expanding the students' experiences to include computer simulations as well as direct, hands-on projects with communications circuits, devices, and test equipment.

Most of the projects included in this manual require the student to interact with simulated equipment circuits, and devices located on the computer screen. Many of these are *prep projects*—projects that prepare the student for more effective use of limited lab time. Every prep project precedes a *hands-on project*. The hands-on projects are to be conducted in the laboratory, using conventional instruments and procedures. (See Appendix A for a complete listing of components, supplies, and equipment required for the hands-on projects.)

In conventional hands-on communications labs, most time is spent gathering the components, wiring the circuits, and adjusting the test instruments. Once the circuit is operating and the test instruments are adjusted properly, little time remains for thoughtful and thorough experimentation. The prep projects in this laboratory system eliminate the need for constructing the circuit and struggling with equipment adjustments. The projects allow the student to concentrate upon the tasks of gathering data and, more important, observing the way a circuit responds to changing input conditions. But having gained this valuable experience with the operating principles of a circuit, the student can then move to the corresponding hands-on lab where he or she can then afford to give full attention to the mechanics of constructing actual circuits and adjusting real instruments.

Few school laboratories can offer ready access to the full range of instruments and circuits that represent the new domains of communications technology. Hands-on projects in such instances are entirely impractical. But the laboratory system you are using here can simulate lab experiences with sophisticated, expensive, and potentially dangerous equipment. This is the purpose of the *extended* projects.

The simulated projects all use dynamic, interactive, graphical formats; the student can adjust input parameters to test the circuits and view the results in real time. Voltage and frequency sources, for example, are adjusted by using the mouse to grasp and slide simulated controls located on an image of the instrument. The output values resond instantly, usually in a digital format on simulated displays. Outputs from the circuit are also are displayed in real time on simulated instruments. These output devices often use a digital format, but analog displays are used frequently enough to help the student feel comfortable with them (and note their advantages under certain types of test conditions).

Yet another category of projects in this laboratory system are built around two pieces of Electronics Workbench application software: multiSIM 2001 and commSIM 2001. These circuit-simulation files are supplied separately on the laboratory CD-ROM. A student version of multiSIM is supplied with this disc. The commSIM 2001 is available from Electronics Workbench. You can learn more about the higher-end multiSIM and the commSIM software at the manufacturer's website:

www.electronicsworkbench.com

The projects in this manual follow closely the material presented in the second edition of Louis E. Frenzel's *Principles of Electronic Communication Systems*. Each lab assignment, for example, references relevant sections of the textbook. Also, the nomenclature, abbreviations and mathematical notations in this manual are consistent with those presented in the textbook. There are far more hours of laboratory work presented here than is normally available in the electronics and computer labs. The instructor can draw upon a selection of hands-on and simulated lab projects deemed most important for supporting the pace and content of the lecture. Also, most simulated labs are suitable for the instructor to demonstrate in classroom.

The author maintains some space on the Web that is devoted to hints and further suggestions. The URL is:

www.sweethaven.com/commlab

A student of communications electronics traditionally sits on a lab stool at a bench on which there are a few pieces of lab equipment and a handful of transistors, ICs, coils, and other small components. With the laboratory system used in this manual and the accompanying CD-ROM, the lab environment is expanded to include the simulation effects of a computer and the vast information resource of the Internet. Welcome to education in the twenty-first century.

David L. Heiserman

PASSIVE FILTERS

A Prep Project

This project is a computer simulation of laboratory tests for determining the frequency response of four basic types of passive filters—low-pass, high-pass, bandpass, and band-reject filters. For each of the circuits you will:

- Calculate the cutoff and center frequencies as appropriate.
- Gather date for the response curves.
- Plot the response curves on semilog graphs.
- Determine the actual cutoff frequencies from the graphs.

Preparation

Read Frenzel, *Principles of Electronic Communication Systems*, Section 2-3.

Setup Procedure

1. Select **Prep Projects** from the **Projects** menu.

2. Select **Project 1   Passive Filters**.

LAB PROCEDURE

For all parts of this project you will be using the simulated function generator and AC voltmeter. The function generator supplies the required input waveform, and the AC voltmeter provides a convenient means for monitoring the output voltage level. The schematic diagrams accurately represent the circuit under test. The block diagrams indicate how the equipment and circuit are interconnected.

PART 1   LOW-PASS *RC* FILTER

1. Calculate the cutoff frequency for the low-pass *RC* filter circuit shown on the screen as the schematic diagram for Part 1. Record the values of *R* and *C*, and your calculated value of f_{co} in Part 1 of the Results Sheet.

2. Adjust the amplitude of the function generator for 10.0 V_{p-p}. Given this 10-V input, calculate the amount of voltage that should appear at the output of the circuit when the frequency is adjusted to f_{co}. Record the calculated value on the Results Sheet.

3. Adjust the frequency of the function generator to each of the values shown in Table 1-1 on the Results Sheet. In each case, record the value of V_{out} as shown on the AC voltmeter.

4. Set the frequency of the function generator to the value of f_{co} that you calculated in Step 1. While observing the output on the AC voltmeter, adjust the frequency slightly above and below the calculated value of f_{co}. Note the frequency value that causes the output voltage to equal the cutoff voltage that you calculated in Step 3. Record this frequency as measured f_{co} on the Results Sheet.

5. Calculate the dB response of the circuit for each of the measurements in Table 1-1, and then plot the data on the semilog graph shown as Figure 1-1 on the Results Sheet. Also include a point for your measured value of f_{co}.

PART 2   HIGH-PASS *RC* FILTER

1. Calculate the cutoff frequency for the high-pass *RC* filter circuit shown on the screen as the schematic diagram for Part 2. Record the values of *R* and *C* and your calculated value of f_{co} on the Results Sheet.

2. Adjust the amplitude of the function generator for 10.0 V_{p-p}. Given this 10-V input, calculate the amount of voltage that should appear at the output of the circuit when the frequency is adjusted to f_{co}. Record the calculated V_{out} on the Results Sheet.

3. Adjust the frequency of the function generator to each of the values shown in Table 1-2 on the Results Sheet. In each case, record the value of V_{out} as shown on the AC voltmeter.

4. Set the frequency of the function generator to the value of f_{co} that you calculated in Step 1. While observing the output on the AC voltmeter, adjust the frequency slightly above and below the calculated value of f_{co}. Note the frequency value that causes the output voltage to equal the cutoff voltage that you calculated in Step 3. Record this frequency as measured f_{co} on the Results Sheet.

5. Calculate the dB response of the circuit for each of the measurements in Table 1-2. Plot this output data on the semilog graph shown in Figure 1-2 on the Results Sheet. Also include a point for your measured value of f_{co}.

PART 3 SERIES *LC* BANDPASS FILTER

1. Note on the screen the schematic diagram for Part 3. Record the values of L and C on the Results Sheet. Calculate and record the following on the Results Sheet: the circuit's center frequency f_c, the lower -3-dB cutoff frequency f_1, and the upper cutoff frequency f_2.

2. Adjust the amplitude of the function generator for 10.0 V_{p-p}. Adjust the frequency of the function generator to each of the values shown in Table 1-3 on the Results Sheet. In each case, record the value of V_{out} as shown on the AC voltmeter.

3. Set the frequency of the function generator to the value of f_c that you calculated in Step 1. While observing the output on the AC voltmeter, adjust the frequency slightly above and below the calculated value of f_c. Note the frequency value that causes the maximum amount of output voltage. Record this frequency on the Results Sheet.

4. Set the frequency of the function generator to the value of f_1 that you calculated in Step 1. While observing the output on the AC voltmeter, adjust the frequency slightly above and below the calculated

value until the output voltage equals 0.707 times the maximum output voltage. Record this measured value of f_1 on the Results Sheet. Repeat the adjustments and measurements for the upper cutoff point f_2.

5. Calculate the dB voltage gain for each of the measurements in Table 1-3. Plot this output data on the semilog graph shown as Figure 1-3 on the Results Sheet. Also include the points for the measured values of f_c, f_1, and f_2 from Steps 3 and 4.

PART 4 SERIES *LC* BAND-REJECT FILTER

1. Record on the Results Sheet the values of L and C shown on the schematic for Part 4. Calculate and record the circuit's values for f_c, f_1, and f_2.

2. Adjust the amplitude of the function generator for 10.0 V_{p-p}. Adjust the frequency of the function generator to each of the values shown in Table 1-4 on the Results Sheet. In each case, record the value of V_{out} as shown on the AC voltmeter.

3. Set the frequency of the function generator to the value of f_c that you calculated in Step 1. While observing the output on the AC voltmeter, adjust the frequency slightly above and below the calculated value of f_c. Note the frequency value that causes the minimum amount of output voltage. Record this frequency on the Results Sheet.

4. Set the frequency of the function generator to the value of f_1 that you calculated in Step 1. Measure the values of f_1 and f_2 as you did in Step 4 of Part 3. Record the findings in Part 4 of the Results Sheet.

5. Calculate the dB voltage gain for each of the measurements in Table 1-4. Plot this output data on the semilog graph shown as Figure 1-4 on the Results Sheet. Also include the points for the measured values of f_c, f_1, and f_2 from Steps 3 and 4.

2

Critical Thinking for Project 1

1. Describe the effect that a shorted capacitor has upon the response curve of an *RC* low-pass filter. Compare this with increasing the value of *C* in an *RC* high-pass filter.

2. Explain why the equations for an *LC* filter are identical to the equations for the resonant frequency of an *LC* circuit.

3. Explain why the circuit for Part 3 would respond as a high-pass filter circuit if the inductor were shorted.

4. The circuit in Part 4 uses a parallel *LC* circuit as a band-reject filter. Sketch an *LC* band-reject filter that uses a series *LC* circuit.

PASSIVE FILTERS

A Hands-On Project

This project deals with the passive versions of four basic filters: low-pass filters, high-pass filters, bandpass filters, and band-reject filters. For each of the circuits in this project you will:

- Construct the circuit.
- Calculate the critical frequencies.
- Gather data for the response curve.
- Plot the response curve on semilog graphs.
- Determine the critical frequencies from the curve.
- Confirm the actual critical frequencies by direct measurement.

Preparation

1. Read Frenzel, *Principles of Electronic Communication Systems*, Section 2-3.

2. Complete the work for Prep Project 1.

Components and Supplies

1	Resistor, 820 Ω
1	Resistor 4.7, kΩ
2	Capacitor, 10 nF
2	Inductor, 1 mH

Equipment

1	Function generator
1	Dual-trace oscilloscope
1	Frequency counter (optional)

The frequency counter is optional because it is possible to use the oscilloscope to determine the circuit's operating frequency.

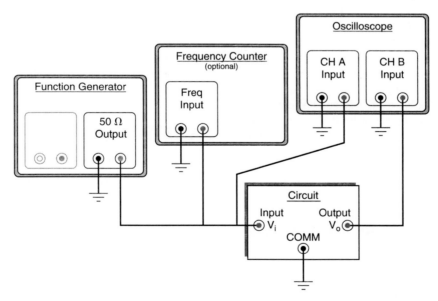

Figure 2-1

LAB PROCEDURE

PART 1   LOW-PASS FILTER

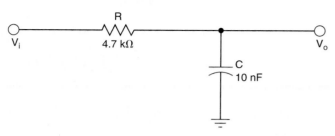

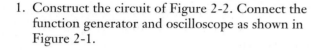

Figure 2-2

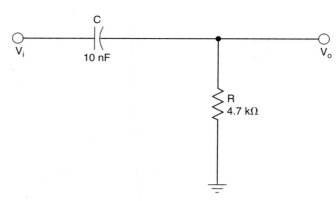

Figure 2-3

1. Construct the circuit of Figure 2-2. Connect the function generator and oscilloscope as shown in Figure 2-1.

2. For each of the entries in Table 2-1 on the Results Sheet, set the function generator for the given frequency and the output voltage to 10 V_{p-p}. Measure and record the peak-to-peak value of V_{out}.

3. Determine the maximum value of V_{out} from the data in Table 2-1. Use this value as the 0-dB level for converting all other values for V_{out} in Table 2-1 to their corresponding dB level. Plot the results on the semilog graph of Figure 2-6.

4. Calculate the cutoff frequency f_{co} according to the values for R and C in Figure 2-2. Estimate the cutoff frequency from the data in your graph. Show your values on the Results Sheet.

5. Determine the −3-dB voltage level for this circuit by multiplying the maximum output voltage by 0.707. Record this value on the Results Sheet. Make sure that V_{in} to the circuit is still 10 V_{p-p}, and adjust the frequency to the actual cutoff frequency (to the point at which V_{out} is the voltage you calculated previously in this step). Record the frequency as the measured cutoff frequency.

PART 2   HIGH-PASS *RC* FILTER

1. Construct the circuit of Figure 2-3. Connect the function generator and oscilloscope as in the previous parts of this project (Figure 2-1).

2. For each of the entries in Table 2-2, set the function generator for the given frequency, double-check the

value of V_{in} to make sure it is 10 V_{p-p}, and record the peak-to-peak value of V_{out}.

3. Determine the maximum value of V_{out} from the data in Table 2-2. Use this value as the 0-dB level for converting all other values for V_{out} in the table to their corresponding dB levels. Plot the results on the semilog graph of Figure 2-7.

4. Calculate the cutoff frequency f_{co} according to the values of R and C in this circuit. Estimate the cutoff frequency from the graph. Record these values on the Results Sheet.

5. Determine the −3-dB voltage level for this circuit as you did for Step 5 of Part 1. Record this value on the Results Sheet. Make sure that the voltage applied at V_{in} is still 10 V_{p-p}, and adjust the frequency to the actual cutoff frequency (to the point at which V_{out} is the voltage you calculated previously in this step). Record the frequency as the measured cutoff frequency.

PART 3   SERIES *LC* BANDPASS FILTER

1. Construct the circuit of Figure 2-4. Connect the function generator, circuit, and oscilloscope as shown for in Figure 2-1.

2. Set the function generator to produce a sinusoidal waveform of 10 V_{p-p} at V_{in}.

3. For each of the entries in Table 2-3, set the function generator for the given frequency, double-check the value of V_{in}, and record the peak-to-peak value of V_{out}.

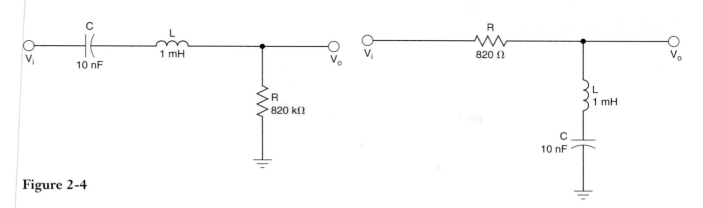

Figure 2-4

Figure 2-5

4. Calculate the center frequency f_c according to the values of L and C given in Figure 2-4. Record your results on the Results Sheet.

5. Set the frequency of the function generator to the calculated value of f_c, and then carefully tune the applied frequency to obtain the maximum level of V_{out}. Record this measured center frequency on the Results Sheet.

6. Convert all values for V_{out} in Table 2-3 to their corresponding dB level. Plot the results on the three-cycle semilog graph of Figure 2-8 on the Results Sheet.

7. Estimate the values of f_1 and f_2 from the graph, and record your values on the Results Sheet.

8. Make sure that V_{in} is still adjusted to 10 V_{p-p}, and determine the measured value of f_c by adjusting the frequency of the function generator to obtain the maximum value of V_{out}. Determine the actual value of f_1 by adjusting the frequency of the function generator below f_c at frequency $0.707 \times V_{in}$, and determine the actual value of f_2 above f_c at frequency $0.707 \times V_{in}$. Record your findings for these measured values on the Results Sheet.

PART 4 SHUNT *LC* BAND-REJECT FILTER

1. Construct the circuit of Figure 2-2 and connect the function generator and oscilloscope as described for Part 1 of this project.

2. For each of the entries in Table 2-4, set the function generator for the given frequency, double-check the 10-V_{p-p} value of V_{in}, and record the peak-to-peak value of V_{out}.

3. Calculate the center notch frequency f_{notch} according to the values of L and C given in Figure 2-5. Record your result on the Results Sheet.

4. Set the frequency of the function generator to the calculated value of f_{notch}, and then fine-tune the applied frequency to obtain the minimum level of V_{out}. Record this measured center frequency and the maximum level of output voltage on the Results Sheet.

5. Convert all values for V_{out} in Table 2-4 to their corresponding dB level. Plot the results on the three-cycle semilog graph of Figure 2-9. *Note:* Plot attenuation values that are beyond -10 dB as -10 dB.

6. Estimate the value of f_1 and f_2 from the graph, and record your values on the Results Sheet.

7. Make sure that V_{in} is still adjusted to 10 V_{p-p}, and determine the measured value of f_{notch} by adjusting the frequency of the function generator to obtain the minimum value of V_{out}. Record your findings on the Results Sheet.

8. Determine the actual value of f_1 by adjusting the frequency of the function generator below f_{notch} for the -3-dB value of V_{out}. Likewise, determine the actual value of f_2 by determining the -3-dB output at a frequency above f_{notch}. Record your findings for the measured values of the cutoff frequencies on the Results Sheet.

PROJECT

2

RESULTS SHEET

PART 1 LOW-PASS FILTER

STEP 4

Calculated f_{co} _____

Estimated f_{co} _____

STEP 5

V_{out} at -3 dB $=$ _____

Measured $f_{co} =$ _____

Table 2-1

f (MHz)	v_o (V_{p-p})	v_o/v_i	20 log(v_o/v_i) (dB)
1			
2			
3			
4			
5			
6			
7			
8			
9			
10			

Questions

1. What is the approximate roll-off rate (in dB per decade) for this circuit?

2. What would happen to the value of f_{co} if the value of C were increased?

3. What would happen to the actual value of f_{co} if the value of V_{in} were decreased?

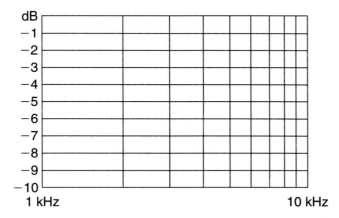

Figure 2-6

PART 2   HIGH-PASS *RC* FILTER

STEP 4

Calculated f_{co} _____

Estimated f_{co} _____

STEP 5

V_{out} at -3 dB = _____

Measured f_{co} = _____

Questions

1. What is the approximate roll-off rate (in dB per decade) for this circuit?

2. What would happen to the value of f_{co} if the value of *C* were increased?

Table 2-2

f (MHz)	v_o (V_{p-p})	v_o/v_i	$20 \log(v_o/v_i)$ (dB)
1			
2			
3			
4			
5			
6			
7			
8			
9			
10			

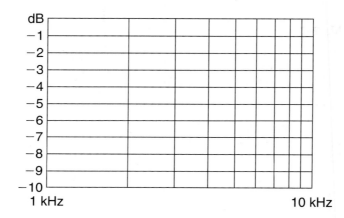

Figure 2-7

PART 3   SERIES *LC* BANDPASS FILTER

STEP 4

Calculated f_c = _____

STEP 5

Measured f_c = _____

STEP 7

Estimated f_1 = _____

Estimated f_2 = _____

STEP 8

Measured f_1 = _____

Measured f_2 = _____

Questions

1. What is the actual Q of this circuit?

2. What is the actual bandwidth of this circuit?

3. Describe any differences you find between calculated and measured values for f_1, f_2, and f_c. Explain your findings.

Table 2-3

f	v_o (V_{p-p})	v_o/v_i	$20 \log(v_o/v_i)$ (dB)
1.5 kHz			
4.5			
7.5			
10.5			
13.5			
15			
45			
75			
105			
135			
150			
450			
750			
1050			
1350			
15 MHz			
45			
75			
105			
135			
150			

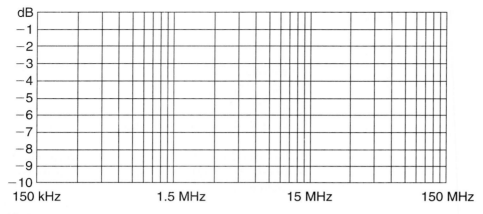

Figure 2-8

PART 4 SHUNT *LC* BAND-REJECT FILTER

Table 2-4

f	v_o (V_{p-p})	v_o/v_i	20 log(v_o/v_i) (dB)
1.5 kHz			
4.5			
7.5			
10.5			
13.5			
15			
45			
75			
105			
135			
150			
450			
750			
1050			
1350			
15 MHz			
45			
75			
105			
135			
150			

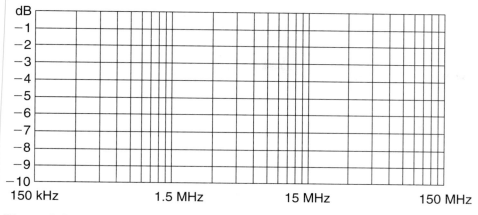

Figure 2-9

STEP 3

Calculated f_{notch} = _____

STEP 4

Measured f_{notch} = _____

Measured $V_{out(min)}$ = _____

STEP 6

Estimated f_1 = _____

Estimated f_2 = _____

STEP 7

Measured f_1 = _____

Measured f_2 = _____

Questions

1. What is the actual Q of this circuit?

2. What is the actual bandwidth of this circuit?

Critical Thinking for Project 2

1. Capacitive coupling is often used between stages of audio amplifiers. Explain why this use causes difficulties with low-frequency audio response.

2. Sketch the schematic diagram of a low-pass LR filter circuit.

3. Suggest a practical reason why RC combinations are more often used than LR combinations in high- and low-pass filter circuits.

4. If the inductor in Figure 2-4 is shorted, the circuit will behave like a different kind of filter circuit. Describe the kind of filter effect that would occur.

5. If the inductor in Figure 2-3 is shorted, the circuit will behave like a different kind of filter circuit. Describe the filter effect that would occur.

PROJECT 3 ACTIVE FILTERS

A Prep Project

This project is a computer simulation of laboratory tests for determining the frequency response of active bandpass and band-reject filters. In each case you will:

- Calculate the center frequency.
- Gather data for the response curve.
- Plot the response curve.
- Determine the center frequency and cutoff frequencies from the curve.
- Confirm the actual center and cutoff frequencies by direct measurement.

Preparation

Read Frenzel, *Principles of Electronic Communication Systems*, Section 2-3.

Setup Procedure

1. Select **Prep Projects** from the **Projects** menu.

2. Select **Project 3   Active Filters.**

LAB PROCEDURE

For both parts of this project you will be using the simulated function generator and AC voltmeter. The function generator supplies the required input waveform. The AC voltmeter provides a convenient means for monitoring the output voltage level.

The block diagrams indicate how the equipment and circuits are interconnected. No schematic diagrams are shown for this project because there are so many practical ways to go about performing the work of active filters. The data you gather here can represent many of the active bandpass and band-reject filters described in your textbook and in the hands-on version of Project 4.

PART 1   ACTIVE BANDPASS FILTER

1. Adjust the amplitude of the function generator for 10.0 V_{p-p}. Adjust the frequency of the function generator to each of the values in Table 3-1 on the Results Sheet. In each case, record the value of V_{out} as shown on the AC voltmeter.

2. Plot the output data from Step 1 on the linear graph in Figure 3-1 on the Results Sheet.

3. Estimate the value of f_c from the curve in Figure 3-1 and adjust the frequency of the function generator to that value. While observing the output on the AC voltmeter, adjust the frequency slightly above and below your estimated value of f_c until you find a frequency that causes the *maximum* amount of output voltage. Record this frequency on the Results Sheet.

4. Set the frequency of the function generator to your estimated value for f_1. While observing the output on the AC voltmeter, adjust the frequency slightly above and below the calculated value until the output voltage equals 0.707 times the maximum output voltage.

5. Record the measured value of f_1 on the Results Sheet. Repeat the adjustments and measurements for the upper cutoff point f_2.

PART 2 ACTIVE BAND-REJECT FILTER

1. Adjust the amplitude of the function generator for 10.0 V_{p-p}. Adjust the frequency of the function generator to each of the values shown in Table 3-2 on the Results Sheet. In each case, record the value of V_{out} as shown on the AC voltmeter.

2. Plot the data from Step 1 on the linear graph in Figure 3-2 on the Results Sheet.

3. As in the previous part of this project, estimate the values of f_c, f_1, and f_2 from the curve in Figure 3-2. Then adjust the frequency of the function generator to determine the precise values. (Remember to adjust for a minimum voltage value when verifying the value of f_c for a band-reject filter circuit.) Record your findings on the Results Sheet.

PROJECT
3

RESULTS SHEET

PART 1 ACTIVE BANDPASS FILTER

STEP 3

Measured f_c = _____

STEP 4

Measured f_1 = _____

Measured f_2 = _____

Questions

1. What is the voltage gain (in V_{out}/V_{in}) of this circuit at the true center frequency?

2. What is the passband of the response curve?

3. What is the value of Q for this circuit?

Table 3-1

f (kHz)	v_o (V_{p-p})	v_o/v_i	20 log(v_o/v_i) (dB)
400			
420			
440			
460			
480			
500			
520			
540			
560			
580			
600			

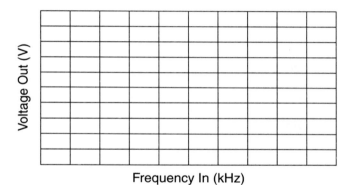

Frequency In (kHz)

Figure 3-1

PART 2   ACTIVE BAND-REJECT FILTER

STEP 3

Measured f_c = _____

STEP 4

Measured f_1 = _____

Measured f_2 = _____

Questions

1. From the data available in Figure 3-2, what is the maximum voltage gain (in V_{out}/V_{in}) for this circuit?

2. What is the bandwidth of the response curve in Figure 3-2?

3. What is the value of Q for this circuit? Show how you arrived at your answer.

Table 3-2

f (kHz)	v_o (V_{p-p})	v_o/v_i	$20 \log(v_o/v_i)$ (dB)
400			
420			
440			
460			
480			
500			
520			
540			
560			
580			
600			

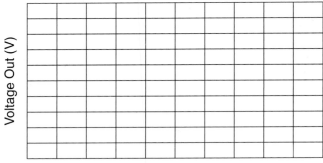

Figure 3-2

Critical Thinking for Project 3

1. Name at least three advantages of active filters over their passive counterparts.

2. Describe the effect that decreasing the values of capacitors will most likely have upon the center frequency of an active bandpass filter circuit.

3. Sketch the circuit for a twin-T version of the active band-pass filter in Part 1.

4. Sketch the circuit for a twin-T version of the active notch filter in Part 2.

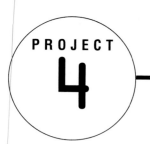

PROJECT 4

ACTIVE FILTERS

A Hands-On Project

In this project you will construct the circuit and determine the frequency response characteristics of active *RC* band-pass and band-reject circuits. For each circuit you will:

- Construct the circuit.
- Calculate the center frequency.
- Gather data for the response curve.
- Plot the response curve on semilog graph.
- Determine the center frequency and cutoff frequencies from the curve.
- Confirm the actual center and cutoff frequencies by direct measurement.

Preparation

1. Read Frenzel, *Principles of Electronic Communication Systems*, Section 2-3.

2. Complete the work for Prep Project 5.

Components and Supplies

1	Resistor, 120 Ω
1	Resistor, 180 Ω
1	Resistor, 470 Ω

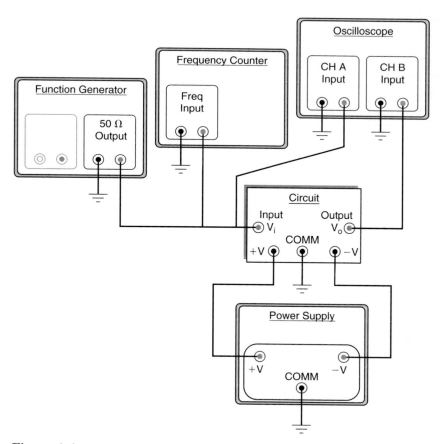

Figure 4-1

1 Resistor, 2.7 kΩ
1 Resistor, 3.3 kΩ
1 Resistor, 22 kΩ
1 Resistor, 330 kΩ
2 Capacitors, 1 nF
1 741 Op amp

Equipment

1 Power supply
1 Function generator
1 Dual-trace oscilloscope
1 Frequency counter

LAB PROCEDURE

PART 1 ACTIVE BANDPASS FILTER

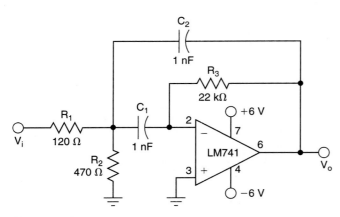

Figure 4-2

1. Construct the circuit of Figure 4-2, and connect the function generator and oscilloscope as shown in Figure 4-1.

2. Set the function generator for a sinusoidal waveform of 100 mV$_{\text{p-p}}$ at V_{in}.

3. Adjust the frequency of the function generator to obtain the maximum level of V_{out}. This will be the center frequency of the response waveform. Record the center frequency and the maximum level of output voltage.

4. Using the results in Step 3, calculate and record the value of V_{out} that should be present when the input frequency is at the lower and upper cutoff frequencies (f_1 and f_2.)

5. Adjust the frequency generator to obtain the values of V_{out} for f_1 and f_2 that you calculated in Step 4. Record these frequencies as the measured cutoff frequencies for this circuit.

6. Use the measured values you have determined for f_1, f_c, and f_2 to sketch an approximation of the response curve for this circuit. Use the semilog graph in Figure 4-4.

PART 2 ACTIVE BAND-REJECT FILTER

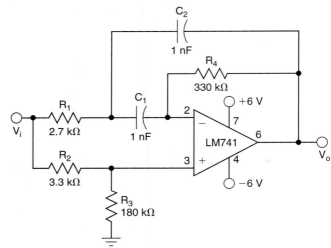

Figure 4-3

1. Construct the circuit shown in Figure 4-3, and connect the function generator and oscilloscope as shown in Figure 4-1.

2. Adjust the function generator for a 500 mV$_{\text{p-p}}$ output, and scan the frequency bands to determine f_{notch}. Record the minimum output voltage level and f_{notch} frequency on the Results Sheet. Also determine the maximum output voltage $V_{\text{out(max)}}$ by tuning the function generator far from the value of f_{notch}. Record this value as well.

3. Calculate and record the half-power voltage level for this circuit as $0.707 \times V_{out(max)}$.

4. Adjust the input frequency to obtain the half-power voltage output level at points below and above f_{notch}. These are the cutoff frequencies (f_1 and f_2) for the circuit. Record the values on the Results Sheet.

5. Adjust the output of the frequency generator to each of the frequencies listed in Table 4-1. In each case, make sure that $V_{in} = 500$ mV, and record the value of V_{out}.

6. Calculate the ratio $V_{out}/V_{out(max)}$ for each measurement, and then convert the result to dB gain. Plot the results on the semilog graph in Figure 4-5.

7. Also plot the values you determined in Steps 2 and 4.

PROJECT **4**

RESULTS SHEET

PART 1 ACTIVE BANDPASS FILTER

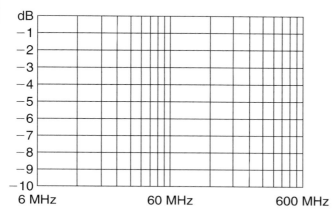

Figure 4-4

STEP 3

$f_c =$ _____

V_{out} at $f_{co} =$ _____

STEP 4

Calculated V_{out} for $f_1 =$ _____

Calculated V_{out} for $f_2 =$ _____

STEP 5

Measured value of $f_1 =$ _____

Measured value of $f_2 =$ _____

Questions

1. What is the voltage gain of the circuit at resonance?

2. What are the bandwidth and Q of the circuit?

PART 2 ACTIVE BAND-REJECT FILTER

Table 4-1

f (kHz)	v_o (V$_{p-p}$)	v_o/v_i	$20 \log(v_o/v_i)$ (dB)
1.0			
2.0			
3.0			
4.0			
5.0			
6.0			
7.0			
8.0			
9.0			
10.0			

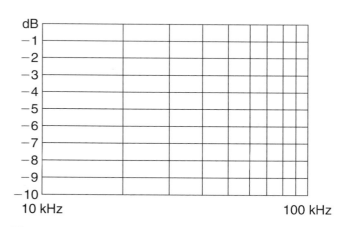

Figure 4-5

STEP 2

f_{notch} = _____

$V_{out(min)}$ = _____

$V_{out(max)}$ = _____

STEP 3

V_{out} at half-power points = _____

STEP 6

f_1 = _____

f_2 = _____

Questions

1. What are the bandwidth and Q of the circuit?

2. What is the voltage gain (V_{out}/V_{in}) at f_{notch} when the input frequency is less than 1 kHz?

Critical Thinking for Project 4

1. Name the components in Figure 4-2 that most likely determine the value of the circuit's resonant frequency.

2. Describe the differences in components and connections between the two circuits in this project. Explain how these few differences account for a great difference in their operating characteristics.

3. Explain how the bandwidth of active filters (such as those in this project) can be decreased.

CERAMIC FILTERS

A Prep Project

This project is a computer simulation of laboratory tests for determining the frequency response of ceramic filters. During the course of this work, you will:

- Gather data for the response curve.
- Plot the response curve.
- Determine the relative amplitude and frequency of multiple response peaks.
- Determine the cutoff frequencies from the curve.
- Confirm the actual center and cutoff frequencies by direct measurement.
- Determine the bandwidth of the circuit.

Preparation

Read Frenzel, *Principles of Electronic Communication Systems*, Section 2-3.

Setup Procedure

1. Select **List Prep Projects** from the **Projects** menu.

2. Select **Project 5   Ceramic Filters.**

LAB PROCEDURE

For both parts of this project you will be using the simulated function generator and AC voltmeter. The function generator supplies the required input waveform. The AC voltmeter provides a convenient means for monitoring the output voltage level.

The schematic and block diagrams are identical for both parts of the project. The only differences between Part 1 and Part 2 are the center frequency and bandwidth of the ceramic filters.

PART 1   455-kHz FILTER

1. Set the output of the function generator to 10 V. For each entry in Table 5-1 on the Results Sheet, set the function generator to the specified frequency and record the resulting peak-to-peak value of V_{out}.

2. Determine the maximum value of V_{out} from the data in Table 5-1. Use this value as the 0-dB level for converting all other values of V_{out} to their corresponding dB level.

3. Plot the values for V_{out} on the linear graph in Figure 5-1. The response curve should have three peaks. From the data you have gathered, cite the frequency and amplitude of each peak. Use the spaces provided for these values on your Results Sheet.

4. From your response curve (see Figure 5-1), determine the upper and lower cutoff frequencies. Record your answers on the Results Sheet.

PART 2   50-MHz FILTER

1. For each of the entries in Table 5-2 on the Results Sheet, set the function generator to the specified frequency and record the resulting peak-to-peak value of V_{out}.

2. Determine the maximum value of V_{out} from the data in Table 5-2. Use this value as the 0-dB level for converting all other values of V_{out} in Table 5-2 to their corresponding dB level.

3. Plot the results on the linear graph in Figure 5-2. The response curve should have five peaks. Indicate on the graph the location of each peak. Estimate the frequency and amplitude of each peak. Verify your estimates by actually tuning the frequency source back and forth across each peak. Record your findings on the Results Sheet.

4. From the response curve in Figure 5-2, determine the upper and lower cutoff frequencies of the filter. Record your answers on the Results Sheet.

PROJECT 5

RESULTS SHEET

PART 1 455-kHz FILTER

STEP 3

Lower-frequency peak:

$f =$ _____ $V =$ _____

Middle-frequency peak:

$f =$ _____ $V =$ _____

Upper-frequency peak:

$f =$ _____ $V =$ _____

Table 5-1

f (kHz)	v_o (V_{p-p})	v_o/v_i	$20 \log(v_o/v_i)$ (dB)
420			
430			
440			
445			
450			
455			
460			
465			
470			
480			
490			

STEP 4

Lower cutoff frequency $f_1 =$ _____

Upper cutoff frequency $f_2 =$ _____

Questions

1. What is the center frequency for the response curve in Figure 5-1? Explain your answer.

2. What is the passband of the response curve in Figure 5-1?

3. What is the Q of the response curve in Figure 5-1?

Figure 5-1

PART 2   50-MHz FILTER

STEP 3

First peak (lowest frequency): $f =$ _____

Second peak: $f =$ _____

Third peak: $f =$ _____

Fourth peak: $f =$ _____

Fifth peak (highest frequency): $f =$ _____

Table 5-2

f (MHz)	v_o (V_{p-p})	v_o/v_i	$20 \log(v_o/v_i)$ (dB)
38			
40			
42			
44			
46			
48			
50			
52			
54			
58			
60			
62			

STEP 4

Lower cutoff frequency $f_1 =$ _____

Upper cutoff frequency $f_2 =$ _____

Questions

1. What is the center frequency for the response curve in Figure 5-2? Explain your answer.

2. What is the passband of the response curve in Figure 5-2?

3. What is the Q of the response curve in Figure 5-2?

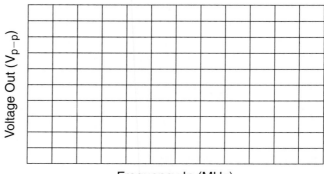

Frequency In (MHz)

Figure 5-2

Critical Thinking for Project 5

1. Describe how the value of Q for a typical ceramic filter compares with the Q for a typical quartz crystal.

2. Explain why a ceramic filter, used alone as in this project, is not considered an active filter.

3. Describe how a ceramic filter might be connected as a band-reject filter.

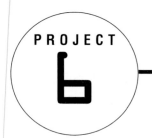

PROJECT 6

CERAMIC FILTERS

A Hands-On Project

In this project you will determine the frequency response characteristics of a 455-kHz ceramic filter. The work requires you to:

- Construct the circuit.
- Gather data for the response curve.
- Plot the response curve.
- Determine the relative amplitude and frequency of multiple response peaks.
- Determine the cutoff frequencies from the curve.
- Confirm the actual peak and cutoff frequencies by direct measurement.

Components and Supplies

2   Resistors, 10 kΩ
1   455-kHz ceramic filter

Equipment

1   Function generator
1   Dual-trace oscilloscope
1   Frequency counter

Preparation

1. Read Frenzel, *Principles of Electronic Communication Systems*, Section 2-3.

2. Complete the work for Prep Project 5.

Ceramic filters often exhibit more than one response peak, the number depending on the manufacturer and the design specifications for the device.

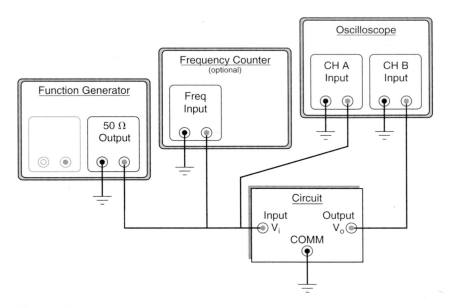

Figure 6-1

       33

LAB PROCEDURE

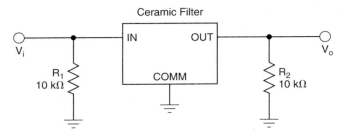

Ceramic Filter

Figure 6-2

1. Construct the circuit of Figure 6-1. Connect the function generator, oscilloscope, and frequency counter as shown in Figure 6-1.

2. Adjust the function generator to obtain a sinusoidal waveform of 10 V_{p-p} at V_{in}.

3. Set the frequency of the function generator to 355 kHz, and then increase the applied frequency while noting the level of V_{out}. Look for more than one response peak at V_{out}. When you find a peak, record the frequency and voltage V_{out} on the Results Sheet.

4. Determine the maximum value of V_{out} from your data in Table 6-1. Use this value as the 0-dB level for determine the value of V_{out} you will find at the -3-dB cutoff frequencies. Record these values on the Results Sheet.

5. Plot the response curve for this device on the graph in Figure 6-3.

PROJECT

6

RESULTS SHEET

STEP 3

Peak 1 frequency = _____

Peak 2 frequency = _____

Peak 3 frequency = _____

STEP 4

Table 6-1

f (kHz)	v_o (V_{p-p})	v_o/v_i	$20 \log(v_o/v_i)$ (dB)
420			
430			
440			
445			
450			
455			
460			
465			
470			
480			
490			

STEP 5

Calculated value of V_{out} at the cutoff frequencies = _____

STEP 6

Measured f_1 = _____

Measured f_2 = _____

Questions

1. How many peaks did you find? What are their frequencies?

2. What is the bandwidth of the circuit?

3. What is the Q of the circuit?

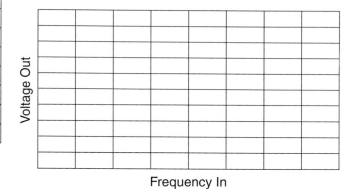

Frequency In

Figure 6-3

Critical Thinking for Project 6

1. Cite some practical advantages of using a ceramic filter as opposed to *LC* or *RC* bandpass filters.

2. What are some disadvantages of using ceramic filters as opposed to *LC* or *RC* bandpass circuits?

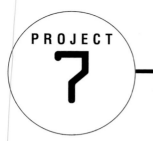

BODE PLOTTER

A multiSIM Project

This project demonstrates the operation of a Bode plotter. The circuit under test is a passive band-pass filter circuit. You will:

- Run the Bode plotter simulation.
- Use the controls on the Bode plotter to determine the three critical frequencies—lower cutoff, center, and upper cutoff frequencies.
- Adjust the values of the capacitor and inductor in the filter circuit and note the responses on the Bode plotter.
- Learn more about using the multiSIM Bode plotter instrument.

Preparation

Read Frenzel, *Principles of Electronic Communication Systems*, Sections 2-1 through 2-3.

Setup Procedure

1. Start multiSIM on your computer.

2. Make sure that your communication lab CD-ROM is in the computer's CD drive.

3. Open the **multiSIM** directory on the CD-ROM.

4. Select **Project_09.msm.**

5. Look for a worksheet diagram that is similar to the one shown in Figure 7-1.

6. Double-click the Bode plotter figure to see the larger functional version (see Figure 7-2).

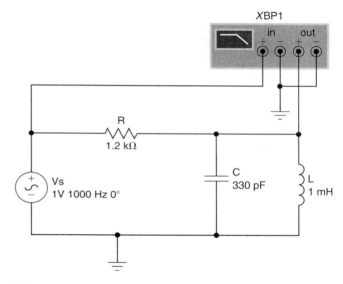

Figure 7-1

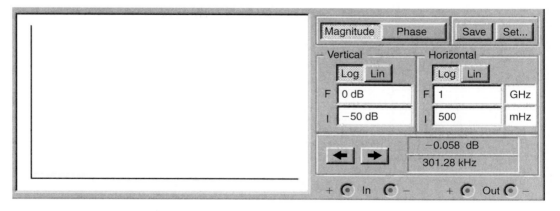

Figure 7-2

LAB PROCEDURE

The Bode plotter is a multiSIM instrument that greatly simplifies the procedures for determining the characteristics of tuned circuits. In effect, the Bode plotter scans a range of frequencies for you and automatically plots the output response.

PART 1 OBSERVE THE OPERATION OF THE BODE PLOTTER INSTRUMENT

1. Run the simulation.

2. Adjust the pointer on the plot to the center of the response curve, and record the center frequency f_c on the Results Sheet.

3. The upper and lower cutoff frequencies are found where the magnitude of the curve is at -3 dB. Adjust the pointer on the plot to determine the upper f_2 and lower f_1 cutoff frequencies. Then determine the bandwidth (BW) of the response. Record this information on the Results Sheet.

4. Sketch the response curve in Figure 7-3. Indicate the locations of the lower cutoff, center, and upper cutoff frequencies.

PART 2 EXPERIMENT WITH THE CIRCUIT VALUES

The capacitor and resistor in this multiSIM setup are both virtual components. This means that you can easily change their values and run the simulation to observe the new response curve.

1. Increase the value of C by a factor of 1000—to 330 nF.

2. Run the simulation.

3. Determine the value of f_c. Record this value on the Results Sheet.

4. Decrease the value of L by a factor of 1000—to 1 μH.

5. Run the simulation.

6. Determine the value of f_c. Record this value on the Results Sheet.

PART 3 LEARN MORE ABOUT USING THE MULTISIM BODE PLOTTER

You can learn a lot more about the Bode plotter by accessing the Help information that is supplied with the multiSIM software.

1. Set the mouse pointer over the display for the Bode plotter instrument.

2. Press the F1 key and the instructions for using the Bode plotter.

Name _____ Date _____

RESULTS SHEET

PART 1 OBSERVE THE OPERATION OF THE BODE PLOTTER INSTRUMENT

STEP 2

f_c = _____ kHz

STEP 3

f_1 = _____ kHz f_2 = _____ kHz

BW = _____ kHz

STEP 4

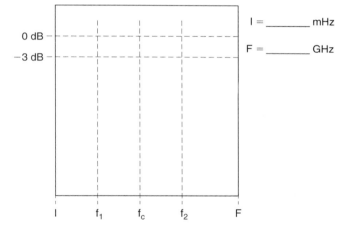

Figure 7-3

PART 2 EXPERIMENT WITH THE CIRCUIT VALUES

STEP 3

f_c = _____ kHz

STEP 6

f_c = _____ kHz

Critical Thinking for Project 7

1. Explain the function of the Log and Lin buttons that appear in the Vertical and Horizontal work boxes on the Bode plotter's large view.

2. Restart multiSIM **Project_9.msm** and click the Phase button. Explain the response curve you see on the Bode plotter.

PROJECT 8

SWEEP GENERATORS

An Extended Project

This project uses a simulated frequency sweep generator and an oscilloscope display to determine the response curve for the most common types of filter circuits. You will:

- Calculate the cutoff frequencies and center frequencies of various types of filter circuits.
- Determine the cutoff frequencies from sweep-generator displays of low- and high-pass filters.
- Determine the cutoff and center frequencies from sweep-generator displays of bandpass and band-reject filters.

Preparation

Read Frenzel, *Principles of Electronic Communication Systems*, Section 2-3.

Setup Procedure

1. Select **Extended Projects** from the **Projects** menu.

2. Select **Project 8   Sweep Generators.**

LAB PROCEDURE

The simulated sweep generator includes the controls required for working with filter circuits. The block diagram shows that outputs of the sweep generator include the sweep frequency (applied at the input of the circuit under test) and a horizontal sawtooth waveform (applied to the horizontal input of the oscilloscope). The output of the circuit under test is applied to the vertical input of the oscilloscope. With this arrangement, the oscilloscope directly displays the amplitude versus frequency response curves you have produced in some of the previous projects.

PART 1   LOW-PASS FILTER

1. Set the amplitude of the sweep generator to 12.6 V. Use this setting for all parts of the project.

2. Set the sweep generator frequency to 0.1 MHz. This is the frequency at which the sweep on the oscilloscope display begins.

3. Set the sweep range to 999.9 MHz. This determines the frequency range of the horizontal sweep on the oscilloscope covers.

   With the frequency set to 0.1 MHz and the sweep range to 999.9 MHz, the sweep on the oscilloscope begins at 0.1 MHz and ends at 0.1 MHz + 999.9 MHz, or 1000 MHz.
   Also notice that there are 10 horizontal divisions on the oscilloscope display. The sweep range is set for 1000 MHz. This means that the horizontal scaling on the oscilloscope is 1000/10, or 100 MHz/div (megahertz per division).

4. Sketch the oscilloscope display on the grid provided as Figure 8-1 on the Results Sheet. Mark the cutoff point on your drawing, and record the cutoff frequency.

   Hint: This waveform has an amplitude that covers three vertical divisions. Therefore, the cutoff amplitude is 0.707 × 3 divisions from the baseline of the waveform.

PART 2   HIGH-PASS FILTER

1. Make sure that the amplitude is still set for 12.6 V.

2. Set the frequency to 0.1 MHz and set the sweep range to 999.9 MHz. Determine the frequency at the left side of the oscilloscope display, the frequency at the right side of the display, and the horizontal scaling. Record your values on the Results Sheet.

3. Sketch the oscilloscope display on the grid provided as Figure 8-2 on the Results Sheet. Mark the point of the cutoff frequency on your drawing and record the value on the Results Sheet.

PART 3   BANDPASS FILTER

1. Make sure that the amplitude is still set for 12.6 V.

2. Set the frequency to 312.0 MHz and set the sweep range to 375.0 MHz. Determine the frequency at the left side of the oscilloscope display, the frequency at the right side of the display, and the horizontal scaling. Record your values on the Results Sheet.

3. Sketch the oscilloscope display on the grid provided as Figure 8-3 on the Results Sheet. On your drawing, mark the points for the center frequency, the upper cutoff frequency, and the lower cutoff frequency. Also record these values on the Results Sheet.

PART 4   BAND-REJECT FILTER

1. Make sure that the amplitude is still set for 12.6 V.

2. Set the frequency to 375.0 MHz and set the sweep range to 250.0 MHz. Determine the frequency at the left side of the oscilloscope display, the frequency at the right side of the display, and the horizontal scaling. Record your values on the Results Sheet.

3. Sketch the oscilloscope display on the grid provided as Figure 8-4 on the Results Sheet. On your drawing, mark the points for the notch frequency, the upper cutoff frequency, and the lower cutoff frequency. Also record these values on the Results Sheet.

PART 5   CERAMIC FILTER

1. Make sure that the amplitude is still set for 12.6 V.

2. Set the frequency to 250.0 MHz and set the sweep range to 500.0 MHz. Determine the frequency at the left side of the oscilloscope display, the frequency at the right side of the display, and the horizontal scaling. Record your values on the Results Sheet.

3. Sketch the oscilloscope display on the grid provided as Figure 8-5 on the Results Sheet. Determine all peak frequencies and the upper and lower cutoff frequencies. Also record these values on the Results Sheet.

RESULTS SHEET

PART 1   LOW-PASS FILTER

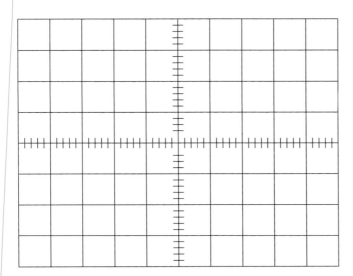

Figure 8-1

STEP 4

  Estimated f_{co} = _____

Questions

1. Assuming that the amplitude of the output waveform is 12.6 V, what is the voltage at the -3-dB level?

2. How many horizontal divisions of this display make up a decade?

PART 2   HIGH-PASS FILTER

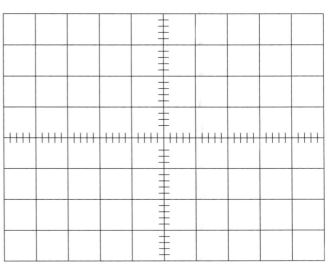

Figure 8-2

STEP 2

  Frequency at start of the sweep = _____

  Frequency at the end of the sweep = _____

  Horizontal scaling of the sweep = _____

STEP 3

  Estimated f_c = _____

  Estimated f_1 = _____

  Estimated f_2 = _____

  Estimated frequencies of the two lower peaks =

  _____ and _____

  Estimated frequencies of the two upper peaks =

  _____ and _____

Results Sheet for Project 8 (*continued*)

Questions

1. Assuming the amplitude of the output waveform is 12.6 V, what is the voltage at the −3-dB level?

2. This waveform is taken from the output of a passive *RC* filter. If the value of *R* is known to be 47 kΩ, what is the value of the capacitor?

PART 3 BANDPASS FILTER

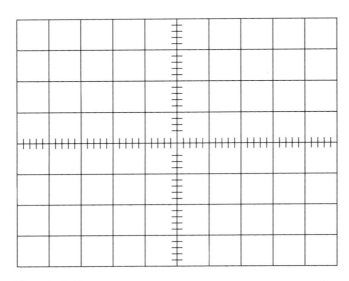

Figure 8-3

STEP 2

Frequency at the start of the sweep = _____

Frequency at the end of the sweep = _____

Horizontal scaling of the sweep = _____

STEP 3

Estimated f_c = _____

Estimated f_1 = _____

Estimated f_2 = _____

Estimated frequencies of the two lower peaks =

_____ and _____

Estimated frequencies of the two upper peaks =

_____ and _____

Questions

1. If the vertical scaling of the display is actually 1 V/div, what is the voltage at the −3-dB level on the waveform?

2. What is the bandwidth of the filter used in Part 3?

PART 4 BAND-REJECT FILTER

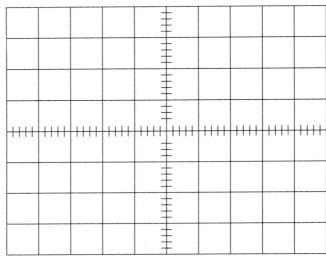

Figure 8-4

STEP 2

Frequency at the start of the sweep = _____

Frequency at the end of the sweep = _____

Horizontal scaling of the sweep = _____

STEP 3

Estimated f_c = _____

Estimated f_1 = _____

Estimated f_2 = _____

Estimated frequencies of the two lower peaks =

_____ and _____

Estimated frequencies of the two upper peaks =

_____ and _____

Questions

1. What is the bandwidth of the filter used in Part 4?

2. With a scaling of 3.2 V/div, what is the voltage at the peak? At the half-power points?

PART 5   CERAMIC FILTER

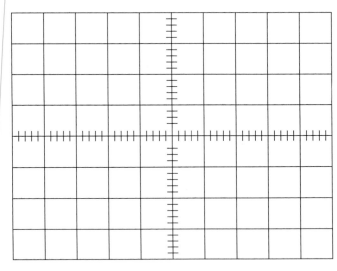

Figure 8-5

STEP 2

Frequency at the start of the sweep = _____

Frequency at the end of the sweep = _____

Horizontal scaling of the sweep = _____

STEP 3

Estimated f_c = _____

Estimated f_1 = _____

Estimated f_2 = _____

Estimated frequencies of the two lower peaks =

_____ and _____

Estimated frequencies of the two upper peaks =

_____ and _____

Questions

1. What is the bandwidth of the filter in Part 5?

2. If the scaling is 1.5 V/div, what is the peak voltage? What is the voltage at the −3-dB points?

Critical Thinking for Project 8

1. The horizontal axis of all the displays in this project is linear. What would be the advantage of using a display that had a semilog scale for its horizontal axis?

2. How do the settings for the frequency and sweep range affect the frequency and amplitude of the half-power points?

3. Explain how a sweep generator is basically a voltage-to-frequency converter that has a sawtooth waveform applied to it.

PROJECT 9

FOURIER ANALYSIS

An Extended Project

This project is a computer simulation of laboratory tests for determining the harmonic frequencies of sinusoidal, square, and triangular waveforms that can pass through a bandpass filter circuit. For each of these waveforms, you will:

- Determine the first through fifth harmonic frequency content.
- Verify the harmonic content by manually tuning an output fifter for peak voltage levels.

Preparation

Read Frenzel, *Principles of Electronic Communication Systems*, Section 2-5.

Setup Procedure

1. Select **Extended Projects** from the **Projects** menu bar.

2. Select **Project 9   Fourier Analysis.**

LAB PROCEDURE

Note the instruments and circuits used in this project. In each part of the project, you will use a function generator to apply the specified waveform shape, amplitude, and frequency to a bandpass filter circuit. You will tune the bandpass filter and note the frequencies that can pass through it. These frequencies are the harmonic content of the waveform. You will use an analog voltmeter to read the amplitude of the waveform from the tuned filter.

PART 1   SINUSOIDAL WAVEFORMS

1. Calculate the frequencies for harmonics 1, 2, 3, 4, and 5 for a sinusoidal waveform of 100 kHz. List your results in the second column of Table 9-1 on the Results Sheet.

2. Set the amplitude of the function generator to its maximum output—49.9 V.

3. Adjust the bandpass filter to each of the five harmonic frequencies you calculated. Determine the corresponding output voltage level for each harmonic, and record the values in the third column of Table 9-1.

PART 2   TRIANGULAR WAVEFORMS

1. Calculate the frequencies for harmonics 1, 2, 3, 4, and 5 for a triangular waveform of 100 kHz. List your results in the first column of Table 9-2 of the Results Sheet.

2. Set the amplitude of the function generator to its maximum output.

3. Adjust the bandpass filter to each of the five harmonic frequencies you calculated. Determine the corresponding output voltage level for each harmonic, and record the values in the second column of Table 9-2.

PART 3   RECTANGULAR WAVEFORMS

1. Calculate the frequencies for harmonics 1, 2, 3, 4, and 5 for a rectangular waveform of 100 kHz. List your results in the first column of Table 9-3 on the Results Sheet.

2. Set the amplitude of the function generator to its maximum output.

3. Adjust the bandpass filter to each of the five harmonic frequencies you calculated. Determine the corresponding output voltage level for each harmonic, and record the values in the second column of Table 9-3.

PROJECT
9

RESULTS SHEET

PART 1 SINUSOIDAL WAVEFORMS

Questions

1. Which harmonics are present, and which are not?

2. If you assign 0 dB to the value of V_{out} at the fundamental frequency, what is the dB gain for each of the harmonics?

Table 9-1

Harmonic	Frequency (Calculated)	v_o (Measured)
First		
Second		
Third		
Fourth		
Fifth		

PART 2 TRIANGULAR WAVEFORMS

Questions

1. Which harmonics are present, and which are not?

2. If you assign 0 dB to the value of V_{out} at the fundamental frequency, what is the dB gain for each of the harmonics?

Table 9-2

Harmonic	Frequency (Calculated)	v_o (Measured)
First		
Second		
Third		
Fourth		
Fifth		

PART 3 RECTANGULAR WAVEFORMS

Questions

1. Which harmonics are present, and which are not?

2. If you assign 0 dB to the value of V_{out} at the fundamental frequency, what is the dB gain for each of the harmonics?

Table 9-3

Harmonic	Frequency (Calculated)	v_o (Measured)
First		
Second		
Third		
Fourth		
Fifth		

Critical Thinking for Project 9

1. Write out the Fourier expressions for the triangular and rectangular waveforms. Circle the factors in the equations that indicate the presence of odd harmonics.

2. Waveforms do not usually have perfect shapes. Explain why you might expect to find some even harmonic content in the analysis of a rectangular waveform that is slightly rounded on the corners.

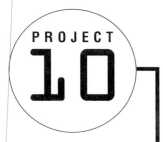

AM SPECTRAL ANALYSIS

PROJECT 10

An Extended Project

This project simulates a test setup that is commonly used for determining the frequency content of AM signals. It uses an instrument called a *spectrum analyzer*. This instrument provides an oscilloscope display that shows the frequency and amplitude of all signals contained within a wide band of frequencies. In this project you will:

- Adjust the frequencies and modulation index for an AM transmitter.
- Use a spectrum analyzer to determine the frequencies present in an AM signal.
- Directly observe how the frequency and amplitudes of the carrier and audio signals at the transmitter affect the content of the broadcast signal.

Preparation

Read Frenzel, *Principles of Electronic Communication Systems*, Section 3-3.

Setup Procedure

1. Select **Projects Extended** from the **Projects** menu.

2. Select **Project 10   AM Spectral Analysis.**

LAB PROCEDURE

In previous AM projects, you used an RF generator and function generator to produce the two signals required for amplitude modulation. The RF generator produced the carrier frequency, and the function generator provided the audio frequency. In this project, these two instruments are combined onto a complete AM generator. With the AM generator, you can adjust the frequency of the carrier and modulation signals. Use the carrier frequency adjustment to determine the carrier frequency, and use the modulation frequency adjustment to set the modulation frequency. The modulation adjustment lets you set the modulation index between 0.0 and 1.2.

1. Set the carrier frequency to 5.0 MHz and the modulation frequency to 1000.0 kHz. Adjust the modulation to 1.0. You are now modulating a 5-MHz carrier with a 1-MHz signal at 100 percent modulation. Use the graph in Figure 10-1 on the Results Sheet to sketch the waveform you see on the spectrum analyzer display. Identify and label the carrier, upper sideband, and lower sideband frequencies.

2. Assume that the display of the spectrum analyzer is calibrated at *5*|MHz/div on the horizontal axis, and 50 V/div on the vertical axis. Record the frequencies and amplitudes of the peaks found on the display.

3. Complete the data in Table 10-1 by setting up the indicated frequencies and modulation indexes, and then recording the spectral response.

RESULTS SHEET

PROJECT 10

STEP 2

Lower sideband:

frequency = _4.0 MHz_

amplitude = _75 V_

Carrier:

frequency = _5.0 MHz_

amplitude = _150 V_

Upper sideband:

frequency = _6.0 MHz_

amplitude = _75 V_

Questions

1. How well does the frequency spectrum of Figure 10-1 line up with the theory of AM sidebands? Explain your answer.

2. How do changes in the carrier frequency affect the spectral display? *Shifts the display left or right*

3. How do changes in the modulation frequency affect the spectral display? *Tightens or expands the sidebands*

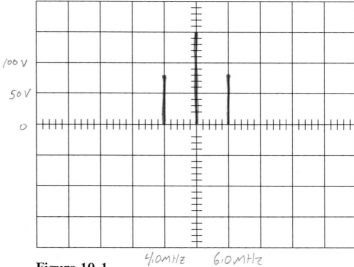

Figure 10-1

Table 10-1

Carrier Freq (MHz)	Modulation Freq (kHz)	Modulation Index	LSB Freq	LSB Ampl	Carrier Freq	Carrier Ampl	ISB Freq	ISB Ampl
2.0	1000.0	1.0	1.0 MHz	75 V	2.0 MHz	150 V	3.0 MHz	75 V
8.0	1000.0	1.0	7.0 MHz	75 V	8.0 MHz	150 V	9.0 MHz	75 V
2.0	500.0	1.0	1.5 MHz	75 V	2.0 MHz	150 V	2.5 MHz	75 V
5.0	500.0	1.0	4.5 MHz	75 V	5.0 MHz	150 V	5.5 MHz	75 V
8.0	500.0	1.0	7.5 MHz	75 V	8.0 MHz	150 V	8.5 MHz	75 V
2.0	1000.0	0.5	1.0 MHz	40 V	2.0 MHz	150 V	3.0 MHz	40 V
5.0	1000.0	0.5	4.0 MHz	40 V	5.0 MHz	150 V	6.0 MHz	40 V
8.0	1000.0	0.5	7.0 MHz	40 V	8.0 MHz	150 V	9.0 MHz	40 V
5.0	1000.0	0.0	4.0 MHz	0 V	5.0 MHz	150 V	6.0 MHz	0 V
5.0	1000.0	1.2	4.0 MHz	95 V	5.0 MHz	150 V	6.0 MHz	95 V

Critical Thinking for Project 10

1. Describe how changes in the modulation index affect the spectral display for an AM broadcast signal.

   It varies the sideband amplitude

2. Describe the appearance of the spectral display if the modulation signal is entirely removed.

   No sidebands

3. Describe the appearance of the spectral display if the carrier frequency is not being generated. Explain your answer.

   No display at all since the sidebands are generated by the carrier frequency (i.e. no f_c = no sidebands)

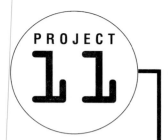

AM MODULATION INDEX

A multiSIM Project

This project demonstrates how an AM waveform is affected by the modulation index. You will:

- Observe the AM waveform on an oscilloscope.
- Vary the modulation index and note the results on the AM waveform.
- Observe the appearance of normal and overmodulated AM waveforms.

Preparation

Read Frenzel, *Principles of Electronic Communication Systems*, Section 3-2.

Setup Procedure

1. Start multiSIM on your computer.

2. Make sure that your communication lab CD-ROM is in the computer's CD drive.

3. Open the **multiSIM** directory on the CD-ROM.

4. Select **Project_11.msm.**

5. Look for a worksheet diagram that is similar to the one shown in Figure 11-1.

LAB PROCEDURE

This simulation uses an oscilloscope instrument to observe the output of a simple AM device. The carrier frequency is fixed for this project at 10 kHz and 1 V_{peak}. The modulating signal is fixed at 100 Hz, but you will be varying the amplitude of the modulating signal as you change the values for the modulation index.

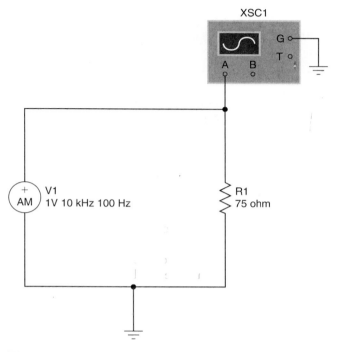

Figure 11-1

PART 1 MODULATION INDEX = 0

1. Make sure that the simulation is not running, and double-click the AM source. You should see the dialog box for the AM Source.

2. Set the **Modulation index (M)** on the dialog box to 0.

3. Close the dialog box.

4. Double-click the oscilloscope instrument to see the expanded view.

5. Start the simulation.

6. Observe the waveform on the oscilloscope display, and adjust the vertical scale for 1 V/div.

55

7. Estimate the value of $V_{max(p-p)}$ from the display and enter the value on the first line in Table 11-1. Also sketch the waveform in Figure 11-2.

PART 2 MODULATION INDEX GREATER THAN 0, LESS THAN 1

1. Stop the simulation and open the AM source dialog box.

2. Set the modulation index to 0.25. Close the dialog box.

3. Start the simulation and check the oscilloscope display.

4. Estimate the values of $V_{max(p-p)}$ and of $V_{min(p-p)}$ and enter them on the corresponding line in Table 11-1.

5. Sketch the waveform in Figure 11-3.

6. Repeat Steps 1 through 4, setting the modulation index to 0.5 and then to 0.75. Record the values of $V_{max(p-p)}$ and of $V_{min(p-p)}$ in the table.

PART 3 MODULATION INDEX EQUALS 1

1. Stop the simulation, open the AM source dialog box, and set the modulation index to 1.

2. Close the dialog box and start the simulation.

3. Estimate the values of $V_{max(p-p)}$ and of $V_{min(p-p)}$ and enter them on the corresponding line in Table 11-1.

4. Sketch the waveform in Figure 11-4.

PART 4 MODULATION INDEX GREATER THAN 1

Repeat the basic steps of this project, setting the modulation index to 2, 3, 4, and 5. Record the value for $V_{max(p-p)}$ in the table, and sketch the waveform for $m = 2$ and $m = 5$ in Figures 11-5 and 11-6, respectively.

Name _____ Date _____

PROJECT
11

RESULTS SHEET

Table 11-1

Modulation Index	$V_{max(p-p)}$	$V_{min(p-p)}$
0		
.25		
.5		
.75		
1		
2		
3		
4		
5		

PART 1   MODULATION INDEX = 0

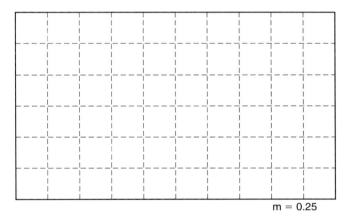

m = 0.25

Figure 11-2

PART 2   MODULATION INDEX GREATER THAN 0, LESS THAN 1

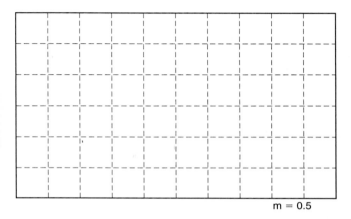

m = 0.5

Figure 11-3

PART 3   MODULATION INDEX EQUALS 1

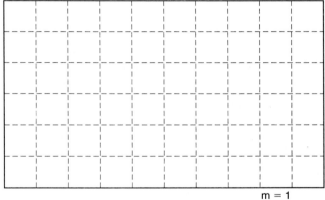

m = 1

Figure 11-4

PART 4   MODULATION INDEX
GREATER THAN 1

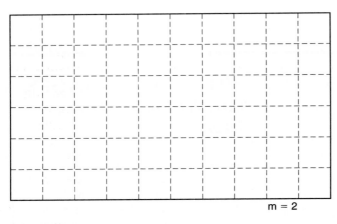

m = 2

Figure 11-5

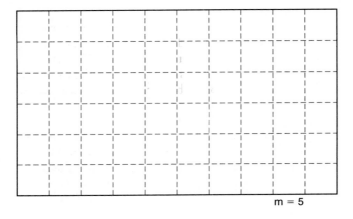

m = 5

Figure 11-6

Questions

1. The AM waveform you show in Figure 11-2 represents a modulation index of 0. What is the voltage level for the modulating signal in this example?

2. Which one of your figures represents 100 percent modulation?

3. Is there any apparent difference in the frequency of the modulation envelope in Figures 11-4 ($m = 1$) and 11-6 ($m = 5$)?

Critical Thinking for Project 11

1. Why do federal regulations prohibit AM signals that have a modulation index greater than 1?

2. A simple mathematical formula shows percent of modulation as a function of modulation index.

Complete the following:

Percent modulation = _____

PROJECT 12

AMPLITUDE MODULATION I

A Prep Project

In this project you will perform computer simulations of tests on a diode amplitude modulator circuit. You will observe and adjust the percent modulation of an AM envelope. You will:

- Sketch and identify the major components of an AM envelope.
- Adjust the AM output for designated levels of modulation.
- Determine the level of modulation from a given modulation waveform.

Preparation

Read Frenzel, *Principles of Electronic Communication Systems*, Section 4-2.

Setup Procedure

1. Select **Prep Projects** from the **Projects** menu.

2. Select **Project 12   Amplitude Modulation I.**

LAB PROCEDURE

This project uses simulated versions of a function generator, an RF generator, and an oscilloscope display. For the purposes of this project, the RF generator provides the carrier signal, and the function generator supplies the audio modulating signal. The frequencies of the function generator and RF generator are not important for the work in this project, so they are fixed at 400 Hz and 1460 kHz, respectively.

The block diagram shows how these instruments are interconnected with the circuit. See your textbook for schematic diagrams of simple diode modulator circuits.

1. Set the output of the function generator for 0.0 V, and adjust the output of the RF generator for an output of 15.0 V. Record this voltage on the Results Sheet. This is the value of V_{out} when there is no audio input.

Note: When the RF output is set to 15.0 V and the audio output from the function generator is set to 0.0 V, you can see that the oscilloscope is vertically scaled at 7.5 V/div. Use this value for determining other peak-to-peak values through this project.

2. Adjust the output of the function generator for an output of 15.0 V. Sketch the waveform at V_{out}, using the grid provided on the Results Sheet as Figure 12-1. Determine the percent modulation of this waveform and record the value on the Results Sheet.

3. Make sure that the output of the RF generator is still set for 15.0 V, and then adjust the output of the function generator for an output of 7.5 V. Sketch the waveform at V_{out}, using the grid in Figure 12-2. Determine and record the percent of modulation on the Results Sheet.

4. Making sure that the output of the RF generator is still set for 15.0 V, adjust the function generator for an output of 3.5 V. Sketch the waveform at V_{out}, using the grid in Figure 12-3. Also determine and record the percent of modulation on the Results Sheet.

5. Adjust the output of the RF generator to 10.0 V. Adjust the output of the function generator in order to produce 75 percent modulation. Sketch the waveform on the grid in Figure 12-4. Read the following values from the oscilloscope display: V_{max}, V_{min}, $V_{max(p–p)}$, and $V_{min(p–p)}$.

6. Make sure that the output of the RF generator is still at 10.0 V. Adjust the output of the function generator for 12.5 V. Sketch the waveform at V_{out}, using the grid provided as Figure 12-5. Also determine and record the percent of modulation on the Results Sheet.

Experimental Notes and Calculations

PROJECT
12

RESULTS SHEET

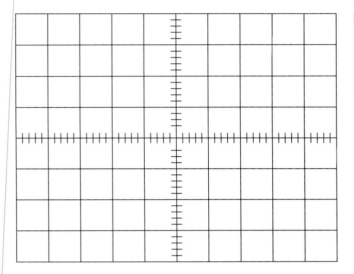

Figure 12-1

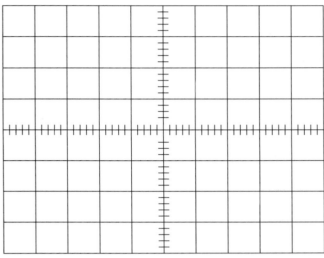

Figure 12-2

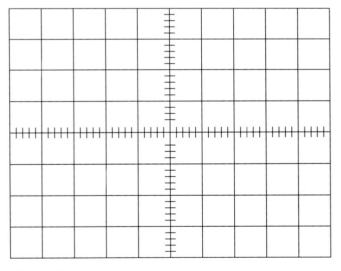

Figure 12-3

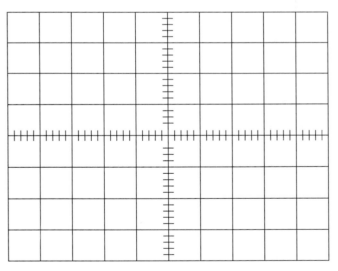

Figure 12-4

STEP 1

V_{out} (audio = 0.0 V) = _____ V_{p-p}

STEP 2

% modulation = _____

STEP 3

% modulation = _____

STEP 4

% modulation = _____

STEP 5

V_{max} = _____ V_{min} = _____

$V_{max(p-p)}$ = _____ $V_{min(p-p)}$ = _____

STEP 6

% modulation = _____

Questions

1. What is the equation for calculating the percent of modulation, given the peak-to-peak values of the carrier signal and the modulating signal?

2. How should you adjust the amplitude settings on the RF generator and function generator in this project to produce 0 percent modulation?

3. Do any of the waveforms on this Results Sheet represent a condition of overmodulation? If so, which one?

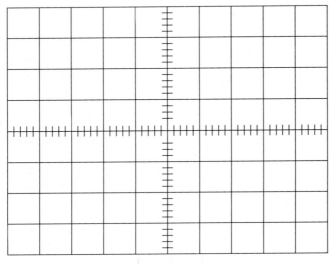

Figure 12-5

Critical Thinking for Project 12

1. Describe the special appearance of a modulation envelope for a signal that is modulated in excess of 100 percent.

2. Sketch the modulation envelopes for a signal that is being modulated with a waveform at 50 percent and at 100 percent.

AMPLITUDE MODULATION I

A Hands-On Project

In this project you will construct and observe the operation of a simple diode AM circuit. The purpose is to acquaint you with the most important features of an AM waveform. Through this work you will:

- Construct the circuit.
- Sketch and identify the major components of an AM envelope.
- Adjust the AM output for designated levels of modulation.
- Determine the level of modulation from a given waveform.
- Sketch modulation waveforms for sinusoidal, triangular, and rectangular audio signals.

Components and Supplies

2	Resistors, 2.2 kΩ
1	Resistor, 100 kΩ
1	Capacitor, 1 μF
1	Capacitor, 10 nF
1	Capacitor, 100 nF
1	Inductor, 1 mH
1	NPN transistor, 2N1904 or equivalent

Equipment

1	Dual-trace oscilloscope
2	Function generators
1	Frequency counter (optional)
1	AM radio receiver

In this project, use one function generator as the audio source, and the second function generator as the carrier source.

Preparation

1. Read Frenzel, *Principles of Electronic Communication Systems*, Section 4-2.

2. Complete the work for Prep Project 12.

LAB PROCEDURE

1. Construct the circuit shown in Figure 13-2. Connect the oscilloscope to output V_{out}, the carrier source (RF generator) to input V_{ic}, and the audio source (function generator) to input V_{ia} (see Figure 13-1).

2. Adjust the carrier source for a 1-V sinusoidal waveform at 50 kHz, and set the audio source for 0 V at 440 Hz. Slightly adjust the frequency of the carrier source for a peak output carrier waveform at V_{out}.

3. Adjust the audio source for a 500-mV output at 440 Hz. Adjust the sweep rate of the oscilloscope until you see a stable modulation envelope.

4. Adjust the amplitude of the audio source to obtain 100 percent modulation at V_{out}. Sketch about three cycles of the waveform in Figure 13-3 on the Results Sheet. Also indicate the value of $V_{max(p-p)}$ on the drawing.

5. Adjust the amplitude of the audio source to obtain 50 percent modulation. Sketch about three cycles of the waveform in Figure 13-4 on the Results Sheet. Also indicate the values of $V_{max(p-p)}$ and $V_{min(p-p)}$ on the drawing.

6. Readjust the amplitude of the audio source to obtain 100 percent modulation. While observing the modulation waveform on the oscilloscope, increase the amplitude of the audio source until you clearly see a waveform that indicates overmodulation. Sketch three cycles of the overmodulation waveform in Figure 13-6 on the Results Sheet. Also indicate the values of $V_{max(p-p)}$ and $V_{min(p-p)}$ on the drawing.

7. With the modulation still set at 100 percent, bring an AM radio receiver close to the circuit. Tune the radio receiver in the 1-MHz (1000-kHz) range until you hear the 440-Hz modulation signal. Tune the receiver for the loudest 440-Hz signal, and note the approximate carrier frequency as determined by

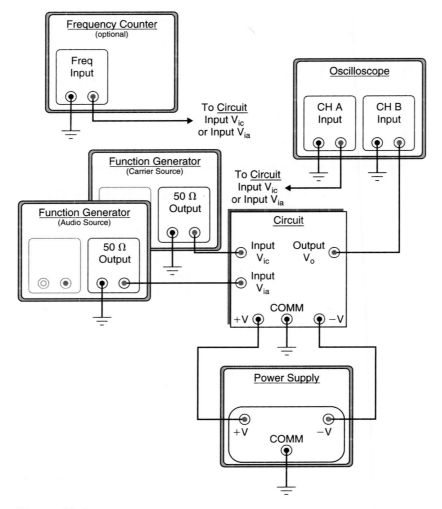

Figure 13-1

reading the tuning dial or LCD display on the receiver. Record this frequency on the Results Sheet.

8. Reduce the level of modulation and note what happens to the sound level from the radio receiver. Note the relationship between percent modulation and the audio signal level from the receiver. Be prepared to describe this effect in the Questions section of the Results Sheet.

9. Switch the audio source to generate a triangular waveform at 1 MHz. Adjust the output level for 100 percent modulation as indicated by the waveform at V_{out}. Sketch three cycles of this modulation waveform in Figure 13-7. Also note the value of $V_{max(p-p)}$ on the drawing.

10. Switch the audio source to generate a rectangular waveform at 1 MHz and adjust the output level for 100 percent modulation. Sketch three cycles of this modulation waveform in Figure 13-8. Also note the value of $V_{max(p-p)}$ on the drawing.

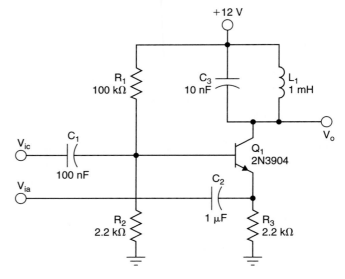

Figure 13-2

PROJECT
13

RESULTS SHEET

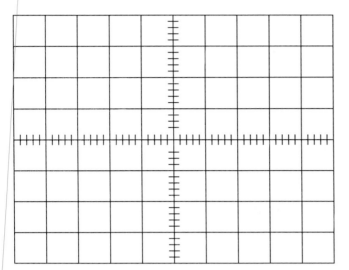

Figure 13-3

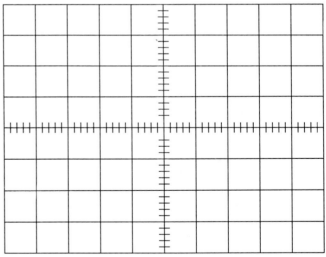

Figure 13-4

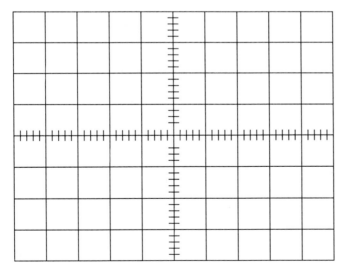

Figure 13-5

Figure 13-6

STEP 7

    Receiver frequency = _____

STEP 8

    Receiver frequency = _____

Questions

1. The voltage levels applied to inputs V_{ic} and V_{ia} often appear greater than the corresponding values that appear on the modulation envelope as V_c and V_m. Explain these differences in amplitude.

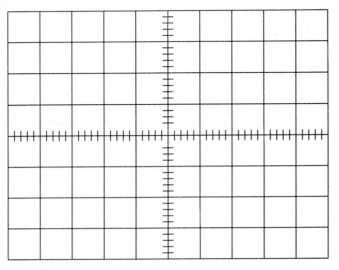

Figure 13-7

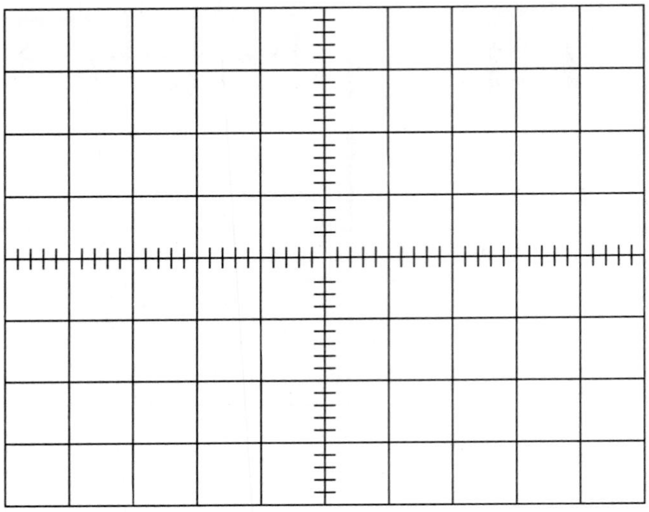

Figure 13-8

2. How would you describe the relationship between percent modulation and the audio signal level from the receiver as noted in Step 8 of this project?

3. What is the *main* purpose of the transistor in this AM modulator circuit?

Critical Thinking for Project 13

1. Explain why capacitor C_2 has a larger value than capacitor C_1.

2. Describe the purpose of the LC resonant circuit and explain why this circuit is important for proper operation of an AM modulator circuit.

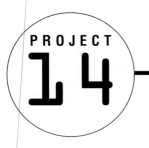

AMPLITUDE MODULATION II

A Prep Project

This project simulates two different setups for reading the modulation level of an AM signal. You have already worked with one of the setups—the time-domain modulation display. This project introduces the trapezoidal display. You will:

- Compare time-domain and trapezoidal displays of typical AM signals.
- Determine the percent of modulation from oscilloscope displays.
- Determine the display that represents a desired level of amplitude modulation.

Preparation

Read Frenzel, *Principles of Electronic Communication Systems*, Section 4-2.

Setup Procedure

1. Select **Prep Projects** from the **Projects** menu.

2. Select **Project 14   Amplitude Modulation II.**

Lab Procedure

This project uses simulated versions of a function generator, an RF generator, and an oscilloscope display. The RF generator provides the carrier signal, and the function generator supplies the audio modulating signal.

Connections to the oscilloscope are different for the time-domain and trapezoidal displays. Clicking the Mode button toggles the system between the two displays. Notice how the block diagrams for the two modes indicate the connections to the oscilloscope.

The formulas for calculating the percent of modulation for the time-domain and trapezoidal displays are essentially the same. The way the measurements are taken are different, however. Clicking the Calc button displays figures that indicate how you should determine the proper values from the oscilloscope.

1. Adjust the amplitude of the RF generator to 30.0 V, and the amplitude of the function generator to 15.0 V. Set the project for the time-domain display, calculate the percent of modulation, and sketch the time-domain waveform in Figure 14-1.

2. Make sure that the signal inputs are at the values used in Step 1. Set the project for the trapezoidal display, calculate the percent of modulation, and sketch the oscilloscope waveform in Figure 14-2.

3. Adjust the amplitude of the RF generator to 30.0 V, and the amplitude of the function generator to 30.0 V. Set the project for the time-domain display, calculate the percent of modulation, and sketch the time-domain waveform in Figure 14-3.

4. Make sure that the signal inputs are at the values used in the previous step. Set the project for the trapezoidal display, calculate the percent of modulation, and sketch the oscilloscope waveform in Figure 14-4.

5. Adjust the amplitude of the RF generator to 15.0 V, and the amplitude of the function generator to 30.0 V. Set the project for the time-domain display, calculate the percent of modulation, and sketch the time-domain waveform in Figure 14-5.

6. Make sure that the signal inputs are at the values used in the previous step. Set the project for the trapezoidal display, calculate the percent of modulation, and sketch the oscilloscope waveform in Figure 14-6.

7. Set the amplitude of the RF generator to 32.0 V. Calculate the amplitude of the audio signal (V_{ia}) required for 80 percent modulation. Record this value of V_{ia} on the Results Sheet and adjust the amplitude of the function generator to that value.

8. Set the project for the time-domain display, sketch the time-domain waveform in Figure 14-7, and determine the values for V_{min} and V_{max} from the display. Record these values on the Results Sheet.

9. Make sure that the signal inputs are at the values used in Step 7. Set the project for the trapezoidal display, sketch the waveform in Figure 14-8, and determine the values for V_{min} and V_{max} from the display. Record these values on the Results Sheet.

PROJECT
14

RESULTS SHEET

STEP 1

  % modulation = _____

STEP 2

  % modulation = _____

STEP 3

  % modulation = _____

STEP 4

  % modulation = _____

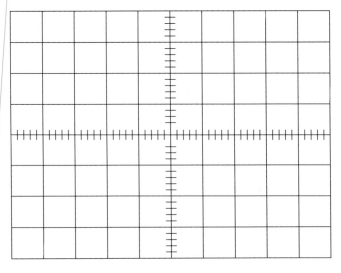

Figure 14-1

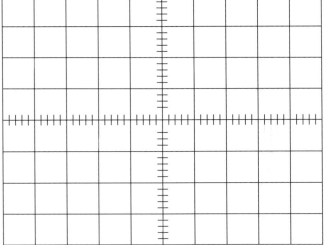

Figure 14-2

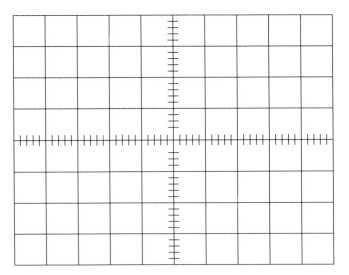

Figure 14-3

Figure 14-4

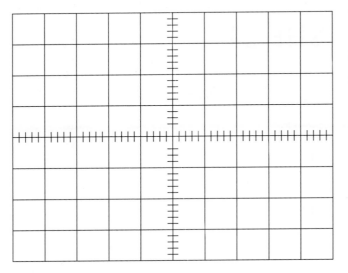

Figure 14-5

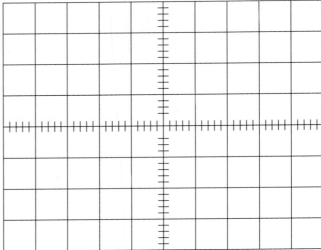

Figure 14-6

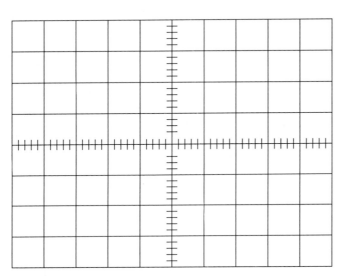

Figure 14-7

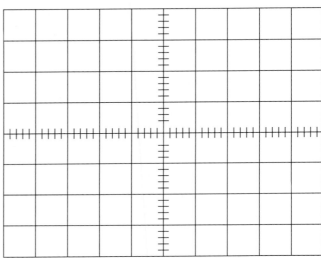

Figure 14-8

STEP 5

    % modulation = _____

STEP 6

    % modulation = _____

STEP 7

    V_{ia} = _____

STEP 8

    V_{min} = _____ V_{max} = _____ %

STEP 9

    V_{ia} = _____

    V_{min} = _____ V_{max} = _____ %

Questions

1. Which figures on this Results Sheet represent modulation levels less than 100 percent?

2. Which figures represent 100 percent modulation?

3. Which figures represent modulation levels greater than 100 percent?

Critical Thinking for Project 14

1. Give verbal descriptions of a trapezoidal waveform at (a) modulation less than 100 percent, (b) modulation at 100 percent, and (c) modulation greater than 100 percent.

2. Describe the direct effect, if any, that changes in the carrier frequency would have on the shape and size of a trapezoidal waveform.

3. This project simulation represents an ideal circuit. In real-life circuits, it is better to determine the modulation index from values of V_{max} and V_{min} on the oscilloscope than from values of V_{ia} and V_{ic} at the inputs. Explain why this is so.

AMPLITUDE MODULATION II

A Hands-On Project

In this project you will demonstrate the use of an operational transconductance amplifier (OTA) as an AM circuit. You will:

• Construct the circuit and null its output.
• Observe and adjust the percent modulation of AM envelope and trapezoidal waveforms.
• Adjust the carrier and audio inputs to achieve a designated level of modulation.
• Note the effects of overmodulation as determined by a trapezoidal waveform.

Components and Supplies

2   Resistors, 10 Ω
1   Resistor, 47 kΩ
1   Resistor, 100 kΩ
1   Operational transconductance amplifier (OTA), CA3080. Manufacturer's information about the CA3080 is available on the Internet.

Data sheet: www.intersil.com/data/fn/fn4/fn475/fn475.pdf.

Applications Notes: www.intersil.com/data/an/an6/an6668/an6668.pdf.

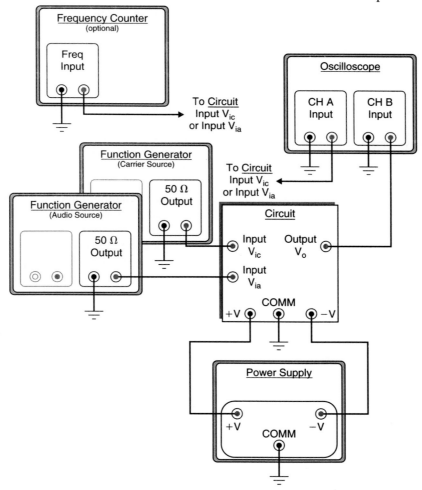

Figure 15-1

Equipment

1   Dual-voltage power supply
1   Dual-trace oscilloscope
2   Function generators
1   Frequency counter (optional)

In both parts of this project, one function generator serves as the audio source and the other as the carrier source.

Preparation

1. Read Frenzel, *Principles of Electronic Communication Systems*, Section 4-2.

2. Complete the work for Prep Project 14.

LAB PROCEDURE

PART 1   TIME-DOMAIN MODULATION DISPLAY

1. Construct the OTA AM circuit shown in Figure 15-2. Connect the oscilloscope V_{out}, the carrier source to V_{ic}, and the audio source V_{ia} (see Figure 15-1).

2. Set the carrier source for a 1-$V_{p–p}$ sinusoidal waveform at 1 MHz, and adjust the audio source for a 1-$V_{p–p}$ sinusoidal waveform at 440 Hz. Adjust the sweep on the oscilloscope to show a stable image of the 440-Hz modulation envelope.

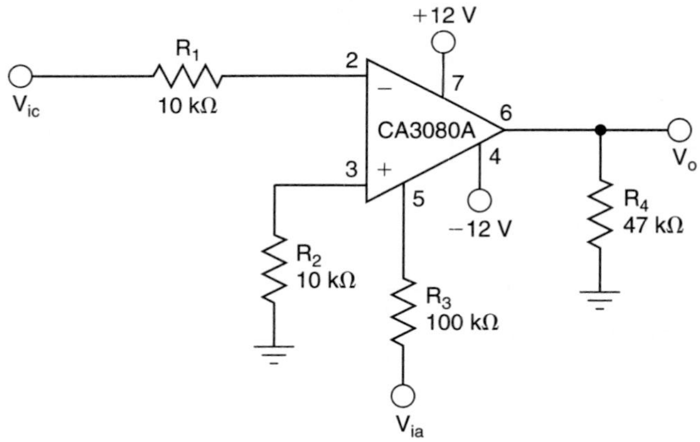

Figure 15-2

3. Adjust the amplitude of the audio source from 0 V to 1.5 V, noting the modulation response on the oscilloscope display. Determine the level of V_{ic} that causes 100 percent modulation. Record this value on the Results Sheet.

PART 2   TRAPEZOIDAL MODULATION DISPLAY

This part of the project requires you to calculate the percent of modulation of an AM envelope from a trapezoidal display. The formula is the same as the formula for determining the percent of modulation from a time-domain AM display. For the trapezoidal display, however, the maximum voltage level refers to the amount of vertical deflection along the right side of the waveform (see the examples in Figure 15-4). The minimum voltage is taken as the amount of vertical deflection along the left side of the display.

1. Set up the oscilloscope for observing the modulated output as a trapezoidal display. To do this, set the oscilloscope for the X-Y mode of operation. Connect Y (vertical) input to V_{out} and the X (horizontal) input to V_{ia}. Both inputs to the oscilloscope should be set for dc input. See Figure 15-3.

2. Adjust the scaling and position of the waveform for a figure that nearly fills the screen. Adjust the amplitude of the audio source for 100 percent modulation. This is indicated by a trapezoidal display such as the one shown in Figure 15-4.

3. Adjust the amplitude of the audio source for 80 percent modulation. Sketch the resulting trapezoidal pattern in Figure 15-5 on the Results Sheet. Also indicate the measured values of V_{max} and V_{min} as determined from the display.

4. Adjust the amplitude of the audio source for 50 percent modulation. Sketch the resulting trapezoidal pattern in Figure 15-6 on the Results Sheet. Also indicate the measured values of V_{max} and V_{min}.

5. Increase the amplitude of the audio source to the point of 100 percent modulation, and then continue increasing the amplitude to about 110 percent modulation. Sketch the resulting trapezoidal pattern in Figure 15-7 on the Results Sheet.

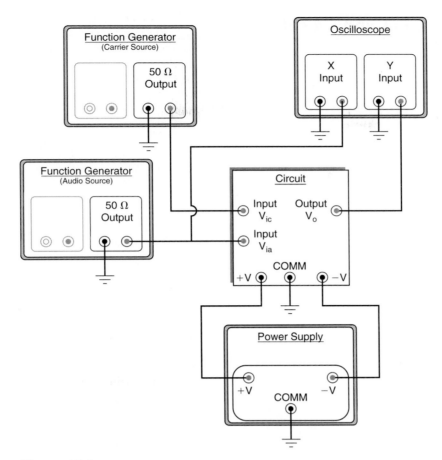

Figure 15-3

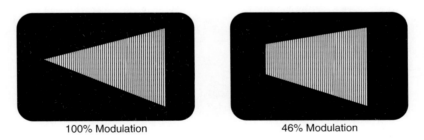

100% Modulation   46% Modulation

Figure 15-4

Experimental Notes and Calculations

RESULTS SHEET

PART 1 TIME-DOMAIN MODULATION DISPLAY

STEP 3

V_{ic} at 100 percent modulation = _____

Questions

1. What is the equation for determining percent modulation as a function of V_{max} and V_{min}?

2. Is input V_{ic} more closely associated with V_{max} or with V_{min} of Question 1? Is input V_{ia} associated with V_{max} or with V_{min}?

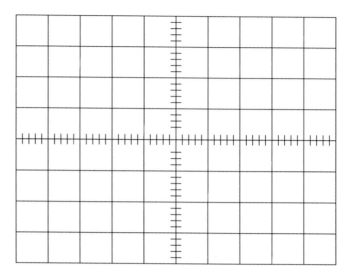

Figure 15-5

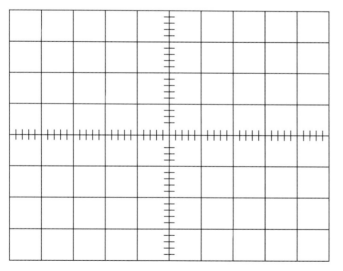

Figure 15-6

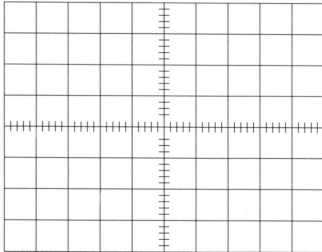

Figure 15-7

PART 2 TRAPEZOIDAL MODULATION DISPLAY

STEP 3

Measured V_{max} = _____

Measured V_{min} = _____

STEP 4

Measured V_{max} = _____

Measured V_{min} = _____

Questions

1. Which terminal on the 3080 OTA device is the inverting input? Noninverting input? Bias?

2. How is overmodulation shown on a trapezoidal display?

Critical Thinking for Project 15

1. Explain why it is better to calculate the modulation index from observed values of V_{max} and V_{min} as viewed on the oscilloscope display than from values of inputs V_{ic} and V_{ia}.

2. Explain why integrated-circuit OTAs are not used for carrier frequencies above 500 kHz.

PROJECT 16

DIODE AM DETECTOR

A Prep Project

In this project you will perform computer simulations of a diode AM detector circuit, noting especially the effects of the output filter capacitor. You will:

- Compare the AM signal before and after detection takes place.
- Sketch waveforms of the signals involved in the AM detection process.
- Observe the purpose of the output filter capacitor.

Preparation

Read Frenzel, *Principles of Electronic Communication Systems*, Section 4-3.

Setup Procedure

1. Select **Prep Projects** from the **Projects** menu.

2. Select **Project 16   Diode AM Detector.**

LAB PROCEDURE

In this project you will use the function generator and RF generator as signal sources for AM modulation and demodulation. The RF generator provides the carrier frequency, and the function generator provides the modulation signal. The schematic diagram shows that you are using a diode modulator and demodulator.

The schematic diagram is an interactive diagram. When you move the mouse cursor over one of the four test points, you will see the corresponding waveform on the oscilloscope display. You can switch the order of the two diagrams by clicking the one you want in the foreground.

1. Set the amplitude of the RF generator to 6.0 V, and the amplitude of the function generator to 4.0 V.

2. Click the schematic diagram to make sure that no part of it is hidden by the block diagram. Move the mouse cursor to TP 1, and sketch the oscilloscope waveform in Figure 16-1 on the Results Sheet. Move the mouse cursor to TP 2, and sketch the waveform on Figure 16-2.

3. Move the mouse cursor to TP 3 and TP 4, and sketch the corresponding waveforms on Figure 16-3.

4. Leave the amplitude of the RF generator to 6.0 V, but adjust the amplitude of the function generator to 8.0 V. Sketch in Figure 16-4 the waveforms you find at TP 3 and TP 4.

5. Leave the amplitude of the RF generator to 6.0 V, but adjust the amplitude of the function generator to 18.0 V. Sketch on Figure 16-5 the waveforms you find at TP 3 and TP 4.

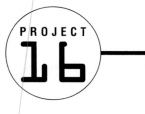

RESULTS SHEET

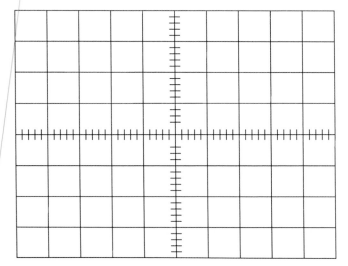

Figure 16-1

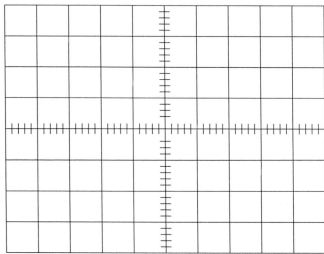

Figure 16-2

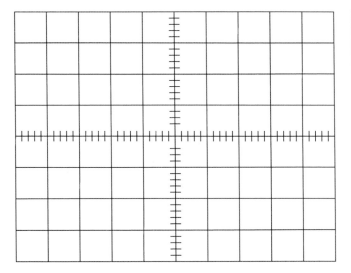

Figure 16-3

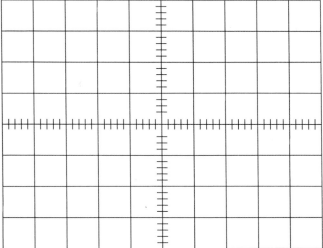

Figure 16-4

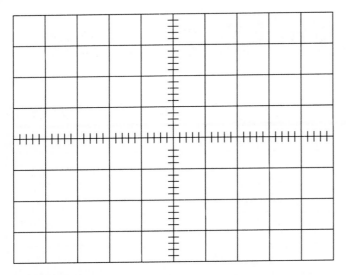

Figure 16-5

Questions

1. What is the V/div scale of the oscilloscope for this project? How did you determine that value?

2. Which, if any, of the figures that you've sketched represents an overmodulated waveform?

Critical Thinking for Project 16

1. Describe the effect that output capacitor C_2 has on the demodulated waveform. Describe how the output waveform would look if C_2 were open.

2. Describe the effect that overmodulation has on the shape of the waveform appearing at the output of the demodulator.

PROJECT 17

DIODE AM DETECTOR

A Hands-On Project

In this project you will study a simple AM diode detector circuit known as an *envelope detector.* You will:

- Construct both an AM modulator and an AM demodulator.
- Compare the signal before and after detection takes place.
- Sketch waveforms of the signals involved in the AM detection process.

Preparation

1. Read Frenzel, *Principles of Electronic Communication Systems,* Section 4-3.

2. Complete the work for Prep Project 16.

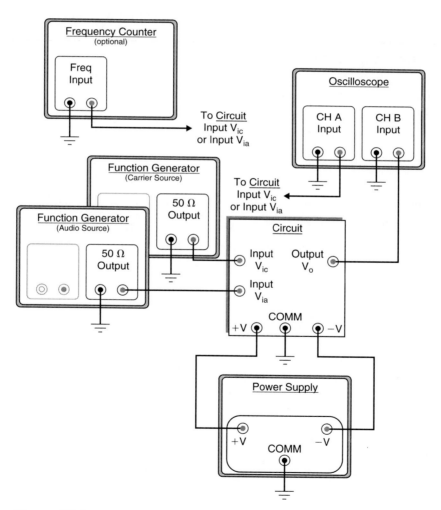

Figure 17-1

Components and Supplies

2	Resistors, 2.2 kΩ
1	Resistor, 27 kΩ
1	Resistor, 100 kΩ
1	Capacitor, 1 μF
1	Capacitor, 10 nF
2	Capacitors, 100 nF
1	Inductor, 1 mH
1	Germanium diode, such as 1N34
1	NPN transistor, 2N2904 or equivalent

Equipment

1	Dual-trace oscilloscope
2	Function generators
1	Frequency counter (optional)

For both parts of this project, use one function generator as the audio source and the other as the carrier source.

LAB PROCEDURE

The circuit for this project is made of the AM modulator circuit in Project 14, followed by a diode detector circuit.

1. Construct the circuit in Figure 17-2. Connect the carrier source to the V_{ic} input, and adjust it for a 1 V_{p-p} sinusoidal waveform at 500 kHz. Connect the oscilloscope to the anode of D_1, and adjust the frequency of the carrier source for a peak voltage response.

2. Connect the audio source to the V_{ia} input, and adjust it for a 400-Hz sinusoidal signal. Set the amplitude of the audio source for 100 percent modulation as seen at the anode of D_1. Sketch at least two cycles of the waveform on the Before axis of the graph in Figure 17-3.

Important: The input to the oscilloscope should be set for DC input for all readings taken from output V_{out}.

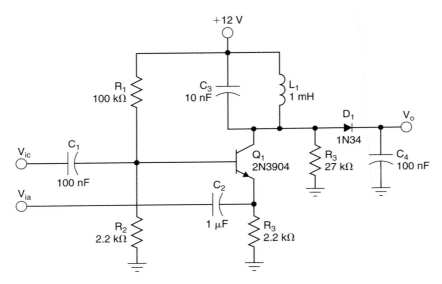

Figure 17-2

3. Connect the oscilloscope to output V_{out}. Sketch at least two cycles of the output waveform on the After axis in Figure 17-3.

4. Adjust the amplitude of the audio source for 80 percent modulation as determined at the modulator output at V_{out}. Sketch at least two cycles of this waveform in Figure 17-4.

5. Adjust the amplitude of the audio source at V_{ia} for 50 percent modulation as determined at the modulator output at V_{out}. Sketch at least two of the output waveforms in Figure 17-5.

6. Set the audio source for a triangular waveform at 400 Hz. Adjust for 100 percent modulation as determined at the modulator output at V_{out}. Sketch at least two cycles of the waveform in Figure 17-6.

7. Set the audio source for a rectangular waveform at 400 Hz and 100 percent modulation as determined at the modulator output at V_{out}. Sketch at least two cycles of the waveform in Figure 17-7.

8. Set the audio source for the sinusoidal output and adjust the amplitude to overmodulate the signal. Set the modulation to a level you estimate to be about 110 percent. Draw at least two cycles of the output waveform in Figure 17-8.

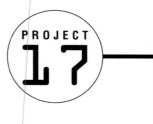

PROJECT 17

RESULTS SHEET

Before

After

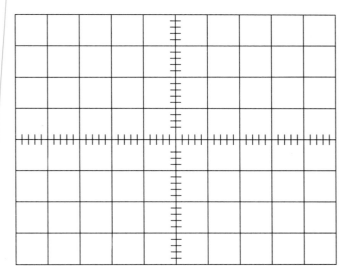

Figure 17-3

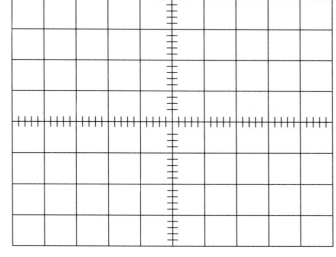

Figure 17-4

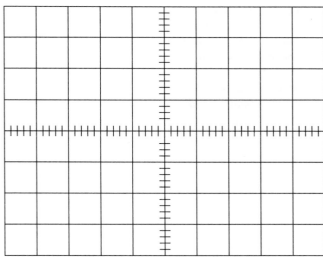

Figure 17-5

Figure 17-6

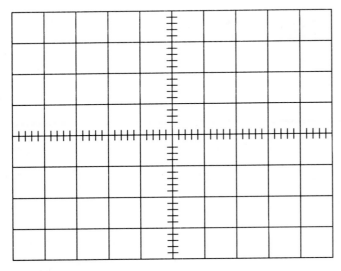

Figure 17-7

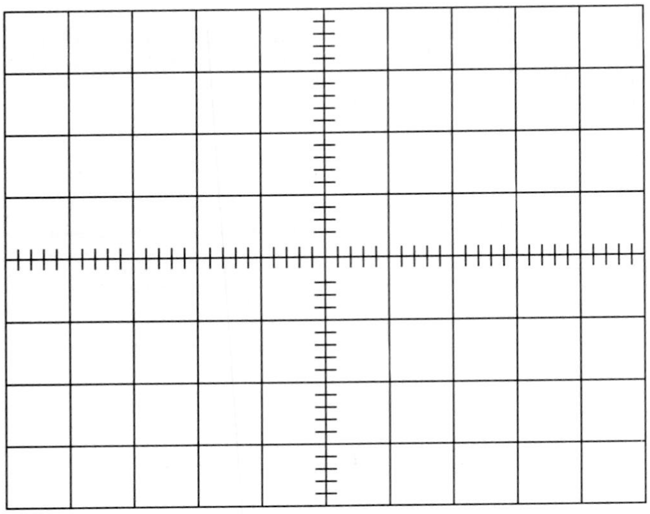

Figure 17-8

Questions

1. Which components make up the demodulator portion of this circuit?

2. How does the percent of modulation affect the amplitude of the output from an AM detector?

Critical Thinking for Project 17

1. Describe how overmodulation affects the appearance (quality) of the signal at the output of an AM demodulator circuit. Explain why overmodulation at a transmitter is undesirable.

2. Name the characteristic that makes a diode useful as an AM detector: rectification, nonlinearity, or both.

3. Explain how reversing the polarity of the detector diode would affect the operation of the detector portion of the circuit.

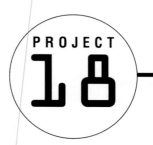
AM DIODE DETECTOR

A multiSIM Project

This project simulates the action of a simple diode detector circuit. You will:

- Observe the proper operation of the circuit in which the modulation index is less than 1.
- Note the distortion of the output waveform when the modulation index is greater than 1.
- Observe the operation of the circuit when the output capacitor is open.

Preparation

Read Frenzel, *Principles of Electronic Communication Systems*, Section 4-3.

Setup Procedure

1. Start multiSIM on your computer.

2. Make sure that your communication lab CD-ROM is in the computer's CD drive.

3. Open the **multiSIM** directory on the CD-ROM.

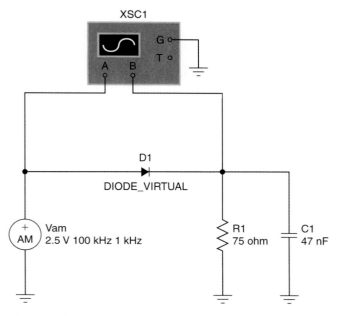

Figure 18-1

4. Select **Project_18.msm**.

5. Look for a worksheet diagram that is similar to the one shown in Figure 18-1.

LAB PROCEDURE

For all parts of this project, V_1 supplies the AM waveform. You will be adjusting the modulation index of V_1 according to the step-by-step lab instructions. The detector diode is, of course, D_1 in the circuit. Resistor R_1 and capacitor C_1 make up the filtering portion of the circuit. In the final part of your lab work for this project, you will be setting the fault parameter for the capacitor.

PART 1   NORMAL OPERATION

1. Open the dialog box for the AM source (V_1), and confirm that it is operating with a modulation index of 0.5.

2. Open the oscilloscope instrument and start the simulation. You will see the AM waveform in red and the demodulated version in blue. Sketch this display on the oscilloscope grid in Figure 18-2.

3. Record the peak-to-peak amplitude of the two waveforms on the Results Sheet.

4. Set the modulation index of the AM source to several other values that are less than 1. Note the resulting waveforms in each case.

Note: You should stop and then restart the simulation after making any change in your values for the modulation index.

PART 2 MODULATION INDEX GREATER THAN 1

1. Open the dialog box for the AM source (V_1), and set the modulation index to 2.

2. Restart the simulation and note the distortion in the demodulated waveform. Sketch this display on the grid in Figure 18-3.

PART 3 OPEN OUTPUT CAPACITOR

The purpose of this part of the project is to illustrate the purpose of the output capacitor C_1.

1. Stop the simulation.

2. Double-click the capacitor device to open its dialog box.

3. Click the Fault tab to show the Fault dialog box.

4. Select the Open fault.

5. Click the open white box for terminal 1. A check mark should then appear in the box.

6. Click the OK button to complete the setup.

7. Set the AM source (V_1) for a modulation index of 0.75.

8. Open the oscilloscope instrument and start the simulation.

9. Sketch the waveforms on the grid in Figure 18-4.

Name _____ Date _____

RESULTS SHEET

PART 1 NORMAL OPERATION

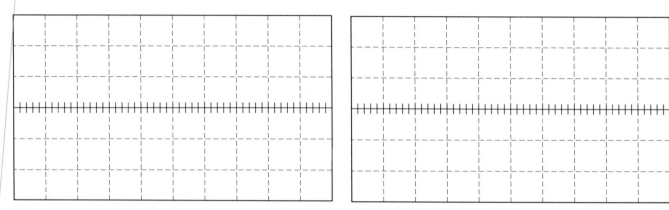

Figure 18-2

STEP 3

　　AM waveform _____ V_{p-p}

　　Demodulated waveform _____ V_{p-p}

Questions

1. What is the peak-to-peak amplitude of the output waveform at 0 percent modulation?

2. What is the value of $V_{min(p-p)}$ at 10 percent modulation for this circuit? At 50 percent modulation?

PART 2 MODULATION INDEX GREATER THAN 1

Figure 18-3

PART 3 OPEN OUTPUT CAPACITOR

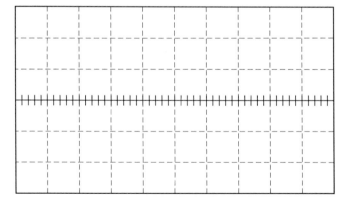

Figure 18-4

Critical Thinking for Project 18

1. Describe the purpose of C_1 in the proper operation of an AM diode detector circuit.

2. What is the purpose of R_1 in the proper operation of this circuit?

3. Compare the operation of this detector with the operation of a half-wave rectifier circuit.

PROJECT 19

SSB MODULATOR

A Prep Project

This project simulates the operation of all four types of AM sideband modulation: double sideband with carrier, double sideband with no carrier, upper single sideband, and lower single sideband. You will:

- Determine the vertical and horizontal scaling of the spectrum analyzer display.
- Adjust the carrier and audio modulating frequencies for all four types of amplitude modulation.
- Read the sideband frequencies and amplitudes from a spectrum analyzer display.

Preparation

Read Frenzel, *Principles of Electronic Communication Systems*, Section 4-5.

Setup Procedure

1. Select **Prep Projects** from the **Projects** menu.

2. Select **Project 19   SSB Modulator.**

LAB PROCEDURE

The function generator provides the audio signal, and the RF generator provides the carrier signal. The spectrum analyzer display provides a convenient and reliable way to determine the frequency and amplitude of the carrier and sidebands.

PART 1   DOUBLE SIDEBAND WITH CARRIER

Note: The purpose of Step 1 is to determine the V/div vertical scaling of the spectrum analyzer display.

1. Make sure that the frequency and amplitude adjustments for the function generator (audio source) are

both set to zero. Set the frequency output of the RF generator to 20.0 MHz. Adjust the amplitude output of the RF generator between its two extremes, and note the carrier frequency amplitude as it appears on the spectrum analyzer display. Determine the amplitude scaling of the display by setting the RF generator amplitude for exactly 1 division. Record this amplitude on the Results Sheet.

Note: The purpose of Step 2 is to determine the hertz per division horizontal scaling of the spectrum analyzer display.

2. Set the output of the function generator to 500 Hz at 8.0 V. Note the upper and lower sidebands, each separated from the carrier frequency by 500 Hz. This separation means that each horizontal division on the display represents 500 Hz.

3. Set up these conditions:

    RF generator frequency = 20.0 MHz
    RF generator amplitude = 10.0 V
    Function generator frequency = 500 Hz
    Function generator amplitude = 10.0 V

Sketch the resulting display on the grid shown in Figure 19-1 on the Results Sheet. Also determine the values requested in that part of the Results Sheet.

4. Set up the following conditions:

    RF generator frequency = 20.0 MHz
    RF generator amplitude = 10.0 V
    Function generator frequency = 750 Hz
    Function generator amplitude = 5.0 V

Sketch the resulting display on Figure 19-2. Also supply the values requested in that part of the Results Sheet.

PART 2   SUPPRESSED-CARRIER DOUBLE SIDEBAND

1. Set up these conditions:

   RF generator frequency = 20.0 MHz
   RF generator amplitude = 10.0 V
   Function generator frequency = 500 Hz
   Function generator amplitude = 10.0 V

Sketch the resulting display on Figure 19-3, and determine the values requested in that part of the Results Sheet.

2. Set up the following conditions:

   RF generator frequency = 20.0 MHz
   RF generator amplitude = 10.0 V
   Function generator frequency = 750 Hz
   Function generator amplitude = 5.0 V

Sketch the resulting display as Figure 19-4, and provide the values requested in that part of the Results Sheet.

PART 3   SUPPRESSED-CARRIER LOWER SIDEBAND

1. Set up the conditions specified earlier in Step 1 of Part 2. Record your results on Figure 19-5 of the Results Sheet.

2. Set up the conditions specified in Step 2 of Part 2. Record your results on Figure 19-6 of the Results Sheet.

PART 4   SUPPRESSED-CARRIER UPPER SIDEBAND

1. Set up the conditions specified earlier in Step 1 of Part 2. Record your results on Figure 19-7 of the Results Sheet.

2. Set up the conditions specified in Step 2 of Part 2. Record your results on Figure 19-8 of the Results Sheet.

PROJECT 19

RESULTS SHEET

Name _____ Date _____

PART 1 DOUBLE SIDEBAND WITH CARRIER

STEP 1

Vertical scale = _____ V/div

STEP 2

Horizontal scale = _____ Hz/div

STEP 3

Carrier amplitude = _____

Carrier frequency = _____

USB amplitude = _____

USB frequency = _____

LSB amplitude = _____

LSB frequency = _____

Modulation index = _____

STEP 4

Carrier amplitude = _____

Carrier frequency = _____

USB amplitude = _____

USB frequency = _____

LSB amplitude = _____

LSB frequency = _____

Modulation index = _____

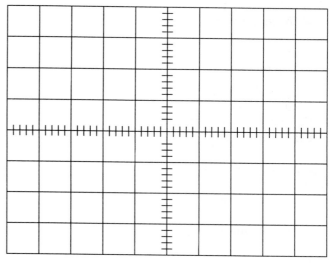

Figure 19-1

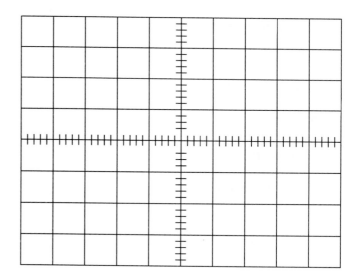

Figure 19-2

Questions

1. What portion of the FM signal are you viewing in Step 1?

2. What is the percent of modulation in Step 3?

3. What is the percent of modulation in Step 4?

PART 2 SUPPRESSED-CARRIER DOUBLE SIDEBAND

STEP 1

    Carrier amplitude = _____

    Carrier frequency = _____

    USB amplitude = _____

    USB frequency = _____

    LSB amplitude = _____

    LSB frequency = _____

    Modulation index = _____

STEP 3

    Carrier amplitude = _____

    Carrier frequency = _____

    USB amplitude = _____

    USB frequency = _____

    LSB amplitude = _____

    LSB frequency = _____

    Modulation index = _____

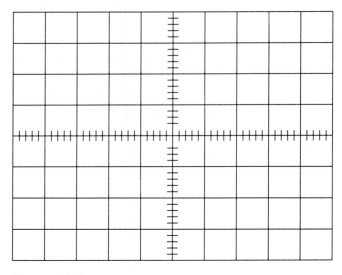

Figure 19-3

Questions

1. What is the main difference between the displays of Part 1 and Part 2?

2. What is the value of the carrier amplitude in Figure 19-3? In Figure 19-4?

3. What is the percent of modulation in Figure 19-4?

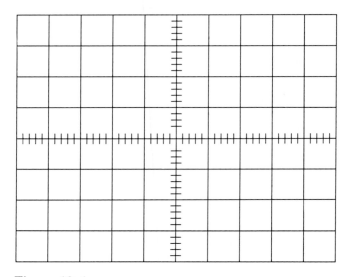

Figure 19-4

PART 3   SUPPRESSED-CARRIER LOWER SIDEBAND

STEP 1

    LSB amplitude = _____

    LSB frequency = _____

    Modulation index = _____

STEP 2

    LSB amplitude = _____

    LSB frequency = _____

    Modulation index = _____

Questions

1. What is the main difference between the displays of Part 2 and Part 3?

2. What is the value of the carrier amplitude in Figure 19-5?

3. What is the value of the USB amplitude in Figure 19-6?

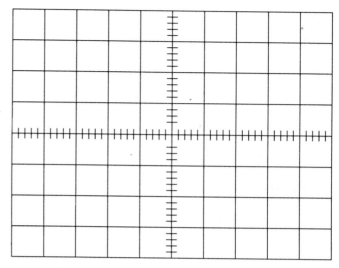

Figure 19-5

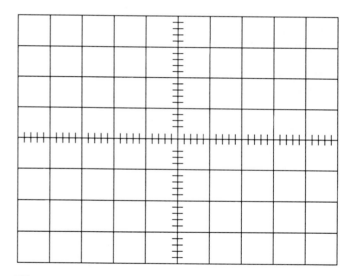

Figure 19-6

PART 4 SUPPRESSED-CARRIER UPPER SIDEBAND

STEP 1

LSB amplitude = _____

LSB frequency = _____

Modulation index = _____

STEP 2

LSB amplitude = _____

LSB frequency = _____

Modulation index = _____

Questions

1. What is the main difference between the displays of Part 3 and Part 4?

2. What are the similarities and the differences between the frequency and amplitude of the peaks in Figures 19-5 and 19-7?

3. What is the value of the LSB amplitude in Figure 19-8?

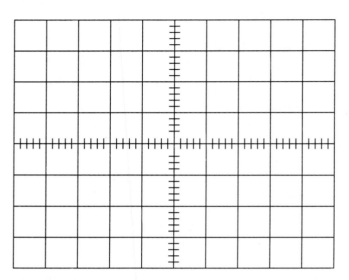

Figure 19-7

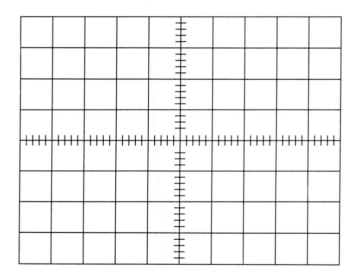

Figure 19-8

Critical Thinking for Project 19

1. Why it is essential to tune an SSB receiver to the same sideband that is being transmitted?

2. Name the type of filtering used in this project and describe an alternative method.

3. Describe the primary advantage of using SSB transmission in the first place.

SSB DEMODULATOR

A Hands-On Project

This project demonstrates how SSB transmissions can be fully demodulated by injecting the carrier frequency at the receiver. You will use a simple shortwave receiver and RF generator to:

- Locate SSB transmissions on the 40- or 10-m bands of a conventional AM shortwave receiver.
- Note the peculiar audio quality of SSB transmission.
- Inject a carrier frequency to produce a satisfactory level of reception.

Preparation

1. Read Frenzel, *Principles of Electronic Communication Systems*, Section 4-5.

2. Complete the work for Prep Project 19.

Equipment

1   Radio receiver capable of receiving the 40- and 10-m shortwave bands. The radio receiver required for this project is commonly included as a feature on consumer CD, tape, or radio appliances or boom boxes.

1   RF generator

LAB PROCEDURE

1. Tune the receiver to the 40-m band (7–7.3 MHz) or 10-m (28–30 MHz) amateur band. Locate an SSB broadcast. On a conventional AM receiver, such broadcasts make a transmitted voice signal sound like someone talking in an empty barrel with a mouth full of marbles.

2. Locate the RF generator directly beside the receiver. Tune the generator through the range of frequencies for the amateur band you have selected on the receiver. Listen for the "beat" frequency (a squealing sound) as the frequency of the generator matches the frequency of the transmission you are receiving.

3. Adjust the generator to obtain the lowest pitch. You should also note that the SSB voice transmission is now more intelligible. Record the generator frequency on the Results Sheet.

4. Select another transmission on the 40- or 10-m band and inject the carrier frequency again. Record the generator frequency for this second transmission.

RESULTS SHEET

STEP 3

Generator frequency = _____

STEP 4

Generator frequency = _____

Questions

1. Assuming the modulating signal is limited to 10 kHz, what is the maximum bandwidth of an SSB transmission?

2. If the modulating signal in Question 1 is transmitted as a DSB-SC signal, what is the maximum bandwidth?

3. Assuming the modulating signal for Step 2 is 0–10 kHz, and that the signal is transmitted as USB-SC, what are the actual minimum and maximum frequencies?

Critical Thinking for Project 20

1. State the primary advantage and disadvantage of SSB transmission and reception.

2. AM receivers that are designed for receiving SSB transmission include a VFO (variable frequency oscillator) control. Explain the purpose of this control and relate it to the work you performed in this project.

LATTICE MODULATOR

A multiSIM Project

Preparation

Read Frenzel, *Principles of Electronic Communication Systems*, Section 4-4.

Setup Procedure

1. Start multiSIM on your computer.

2. Make sure that your communication lab CD-ROM is in the computer's CD drive.

3. Open the **multiSIM** directory on the CD-ROM.

4. Select **Project_21.msm.**

5. Look for a worksheet diagram that is similar to the one shown in Figure 21-1.

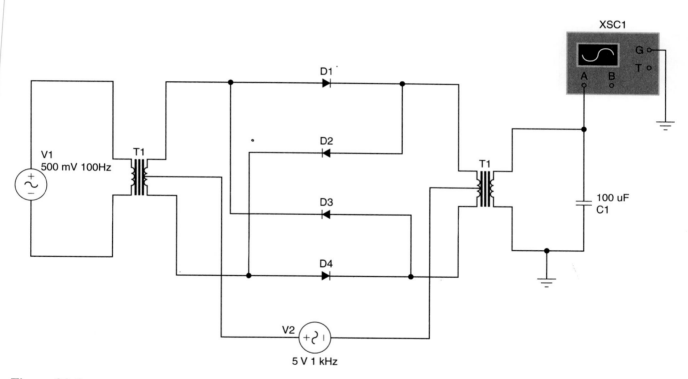

Figure 21-1

LAB PROCEDURE

1. Click the oscilloscope instrument to expand to full size.

2. Start the simulation.

3. Make sure that V_1 is set to 500 mV and V_2 to 5 V. Record $V_{max(p-p)}$ and $V_{min(p-p)}$ on the Results Sheet.

4. Stop the simulation and double-click the V_1 source to show the AC Voltage dialog box.

5. Set the Voltage Amplitude to a level that is necessary for creating 100 percent modulation.

6. Close the dialog box, make sure that the oscilloscope is expanded, and start the simulation.

7. Record $V_{max(p-p)}$ and $V_{min(p-p)}$ on the Results Sheet. Also sketch the waveform in the grid shown as Figure 21-2.

PROJECT
21

RESULTS SHEET

STEP 3

$V_{max(p-p)}$ = _____ $V_{min(p-p)}$ = _____

% modulation = _____

STEP 7

V_1 = _____ V_2 = _____

$V_{max(p-p)}$ = _____ $V_{min(p-p)}$ = _____

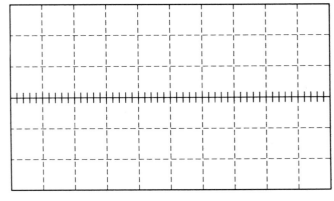

Figure 21-2

Questions

1. Which one of the two voltage sources represents the carrier source? The modulating signal source?

2. Which two diodes conduct when the center tap of T_1 is more positive than the center tap of T_2? When the center tap of T_1 is more negative than the center tap of T_2?

Critical Thinking for Project 21

1. Describe how you would modify the circuit in Figure 21-1 to make it an SSB modulator.

2. Describe the appearance of the output waveform if one of the diodes or transformer winding is open.

FM SPECTRUM ANALYSIS

An Extended Project

This project simulates a test setup for directly observing the carrier frequency and sidebands of an FM broadcast signal. You will:

- Calculate the modulation index.
- Observe the carrier and sideband frequencies
- Verify the Bessel function table.
- Determine the bandwidth.

This computer simulation solves the Bessel functions that describe the frequency spectrum of an FM broadcast signal.

Preparation

Read Frenzel, *Principles of Electronic Communication Systems*, Section 5-3.

Setup Procedure

1. Select **Extended** projects from the Projects menu.

2. Select **Project 22   FM Spectrum Analysis.**

LAB PROCEDURE

The instrument shown as an FM generator gives you complete control over the FM signal. The oscilloscope display for a spectrum analyzer shows the sideband frequencies and their amplitudes.

The two controls on the FM generator allow you to adjust the signal's frequency deviation and carrier frequency. While experimenting with these adjustments, you might discover that they interact with one another under certain circumstances. Actually, the controls are designed so that the setting for the frequency deviation can be no more than four times the setting for the carrier frequency. In other words, the instrument is fixed so that the modulation index of the FM signal cannot exceed 4.

The spectral display includes digital readouts for frequency and amplitude. To use these readouts, move the mouse pointer to the place on the screen where you want to determine frequency and amplitude. Those values automatically appear on the digital readouts.

1. Set the frequency deviation to 0.00 MHz, and the carrier frequency to 125.00 MHz. Sketch the spectrum analyzer display in Figure 22-1 on the Results Sheet. Move the mouse pointer to the peak frequency on the display and record the peak amplitude and frequency. Calculate and record the modulation index for this signal.

2. Set the frequency deviation to 150.00 MHz, and the carrier frequency to 150.00 MHz. Sketch the spectrum analyzer display in Figure 22-2 on the Results Sheet. Determine the amplitude and frequency of each peak, and indicate your values on the figure. Calculate and record the modulation index.

3. Set the frequency deviation to 300.00 MHz, and the carrier frequency to 100.00 MHz. Sketch the spectrum analyzer display in Figure 22-3 on the Results Sheet. Determine the amplitude and frequency of each peak, and indicate your values on the figure. Calculate and record the modulation index.

4. Set the frequency deviation to 1200.00 MHz, and the carrier frequency to 300.00 MHz. Sketch the spectrum analyzer display in Figure 22-4 on the Results Sheet. Determine the amplitude and frequency of each peak, and indicate your values on the figure. Calculate and record the modulation index.

RESULTS SHEET

STEP 1

    Peak voltage = _____

    Peak frequency = _____

    Modulation index = _____

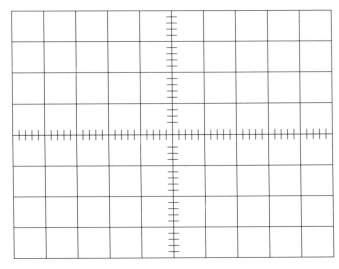

Figure 22-1

STEP 2

    Modulation index = _____

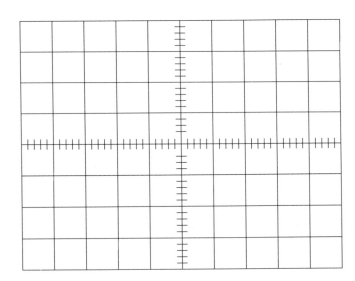

Figure 22-2

STEP 3

Modulation index = _____

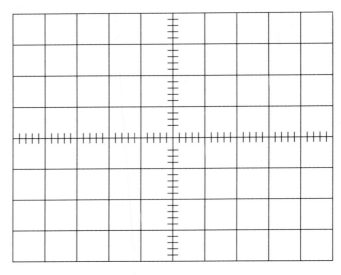

Figure 22-3

STEP 4

Modulation index = _____

Questions

1. According to your table of Bessel functions, how many different frequencies should be present for each of the modulation indexes used in this project?

2. For a signal of the same modulation index how well does each of the voltage levels in Figure 22-4 compare with the voltages cited in your table of Bessel functions?

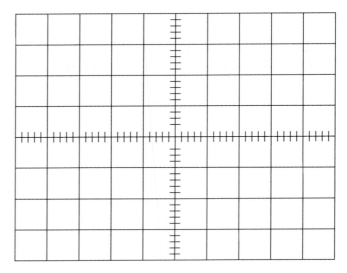

Figure 22-4

Critical Thinking for Project 22

1. Explain the meaning of FM sideband and carrier peaks that have negative values.

2. The modulation index determines the number of significant sidebands. Explain the factors that determine the frequency difference between the individual sidebands.

3. Describe how you could use the simulated instruments of this project to verify the Bessel function curves shown in your textbook.

FREQUENCY-SHIFT KEYING

A multisim Project

Frequency-shift keying (FSK) is a technique for translating binary logic levels into frequency levels. It is a type of frequency modulation that is limited to two voltage-input levels and two frequency-output levels. In this project, you will:

- Observe the operation of a simulated FSK modulator.
- Confirm the voltage and frequency levels.
- Experiment with various combinations of input levels and output frequencies.

Preparation

Read Frenzel, *Principles of Electronic Communication Systems*, Section 5-1.

Setup Procedure

1. Start multisim on your computer.

2. Make sure that your communication lab CD-ROM is in the computer's CD drive.

3. Open the **multisim** directory on the CD-ROM.

4. Select **Project_23.msm**.

5. Look for a worksheet diagram that is similar to the one shown in Figure 23-1.

LAB PROCEDURE

PART 1   DEFAULT SETTINGS

The default settings are the settings that are in place when you first load the project.

1. Double-click the oscilloscope instrument to expand it.

2. Start the simulation.

3. Record the frequency (f_{in}) and peak-to-peak amplitude (V_{in}) of the default digital signal.

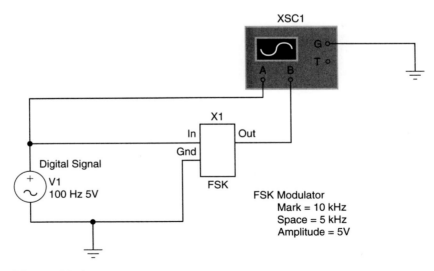

Figure 23-1

---

4. Determine the output frequency (f_{ohi}) when the digital signal is at logic 1, and the output frequency (f_{olo}) when the digital signal is a logic 0.

5. Make a rough sketch of the two waveforms on the grid in Figure 23-2.

PART 2   ADJUSTING THE FSK SETTINGS

1. Turn off the simulation and double-click the FSK block to open the Subcircuit dialog box.

2. Click the Edit Subcircuit button to see the inner workings of the FSK block.

3. Double-click the FSK symbol to see the FSK dialog box, and then select the Value tab.

4. From this box you can adjust the specifications for the FSK—amplitude, mark frequency, and space frequency.

5. Leave the amplitude (A) at 5 V, but adjust the mark (F1) and space (F2) settings to frequencies of your choice. Record your select on the Results Sheet.

6. Close the dialog boxes and run the simulation.

7. Make a rough sketch of the oscilloscope display in Figure 23-3.

Name _____ Date _____

RESULTS SHEET

PART 1 DEFAULT SETTINGS

STEP 3

f_{in} = _____ V_{in} = _____

f_{ohi} = _____ f_{olo} = _____

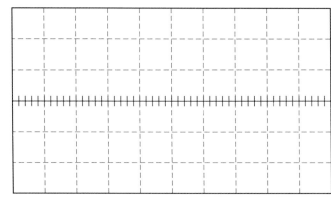

Figure 23-2

PART 2

STEP 5

F1 = _____   F2 = _____

Questions

1. What is the meaning of terms *mark* and space?

2. In this project, does a *mark* represent a logic-0 or a logic-1 level?

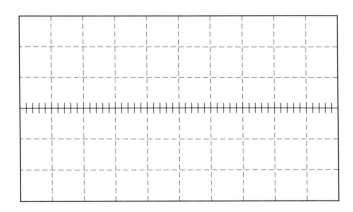

Figure 23-3

Critical Thinking for Project 23

1. Describe the results in this project if the digital signal frequency is equal or greater than the lowest modulation frequency. Explain why the lowest modulation frequency ought be at least ten times greater than the highest digital frequency.

2. According to Fourier theory, the input digital signal is loaded with harmonics. Is this true for the FSK output signal?

113

THE BESSEL FUNCTION BLOCK

A commSIM Project

This project demonstrates the generation of Bessel function curves. You will:

- Observe the generation of the basic family of Bessel curves.
- Adjust and observe the graphical effects of changing sideband numbers *n*.

Preparation

Read Frenzel, *Principles of Electronic Communication Systems*, Section 5-3.

Setup Procedure

1. Start commSIM on your computer.

2. Make sure that your communication lab CD-ROM is in the computer's CD drive.

3. Open the **commSIM** directory on the CD-ROM.

4. Select **Project_24.vsm.**

5. Look for a worksheet diagram that is similar to the one shown in Figure 24-1.

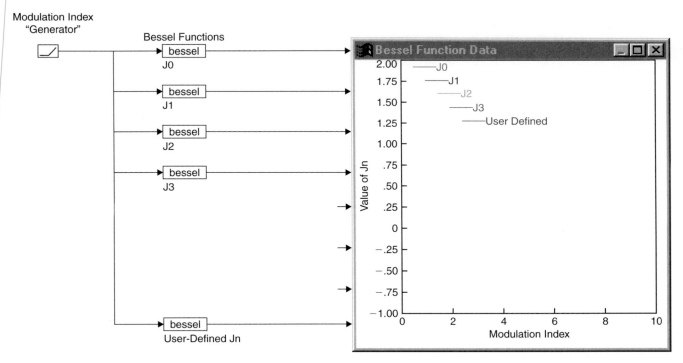

Figure 24-1

115

LAB PROCEDURE

Before starting the simulation, double-click each of the four active bessel blocks to check their parameters. You should find that J0 generates order 0, J1 generates the sidebands of order 1, J2 generates the sidebands of order 2, and so on. The default value for the User-Defined bessel block is set for order 8, but you will be changing that value as part of this project.

Now you are ready to begin the experiment.

1. Start the simulation and compare the resulting curves with the ones shown in your textbook.

2. Sketch waveforms to Figure 24-2.

3. Set the order of the User-Defined Jn to order 6.

4. Restart the simulation and note the curve for J6.

PROJECT
24

RESULTS SHEET

STEP 2

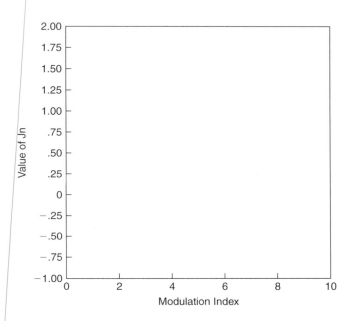

Figure 24-2

Questions

1. Which sideband pair, or order, represents the carrier waveform?

2. At what modulation index does the sideband pair J8 make their largest contribution to the transmitted FM signal?

Critical Thinking for Project 24

1. For an AM signal, the bandwidth is directly related to the frequency deviation of the modulation. Does the same sort of direct relationship exist for FM transmission? Explain your answer.

2. Describe the significance of a value of Jn that is negative, or less than zero.

PROJECT 25

FREQUENCY MODULATION

A Prep Project

In this project you will observe the operation of a circuit that changes DC and AC voltage levels into a corresponding frequency. This is the basis of frequency modulation. You will:

- Observe how the DC input level to a voltage-to-frequency converter affects the output frequency.
- Plot a curve showing output frequency as a function of DC input voltage.
- Observe the operation of a frequency modulator while varying the audio voltage level applied to it.
- Sketch an FM carrier waveform showing the effects of audio modulation.

Preparation

Read Frenzel, *Principles of Electronic Communication Systems*, Section 6-1.

Setup Procedure

1. Select **Prep Projects** from the **Projects** menu.

2. Select **Project 25   Frequency Modulation.**

LAB PROCEDURE

PART 1   DC INPUT

This part of the project uses a DC voltage source to control the frequency of a voltage-to-frequency converter. The oscilloscope display indicates the voltage and frequency levels at the output of the circuit. A digital read-out indicates both the peak-to-peak voltage V_{out} as well as the frequency f_{out}.

1. Adjust the DC voltage source for an output of -10 V (extreme left-hand setting). Then adjust it for an output of $+10$ V (extreme right-hand setting). Note the responses on the oscilloscope display. You do not need to record the responses at this time, however.

2. Set the DC voltage source to each of the voltage levels specified in Table 25-1 on the Results Sheet. Record the corresponding frequencies and voltages that are indicated on the oscilloscope display.

3. From the data of Table 25-1, determine the minimum frequency, maximum frequency, frequency when $V_{in} = 0$, and the maximum frequency deviation. Record your answers on the Results Sheet.

4. From the data of Step 3, calculate the frequency-volt ratio for this circuit. Record your answer on the Results Sheet.

PART 2   AUDIO INPUT

In this part of the project, a function generator applies a sinusoidal AC waveform to the input of the modulation. You are able to vary the amplitude and frequency of that waveform and to note the response on the oscilloscope screen. As indicated on the block diagram for Part 2, the upper trace on the oscilloscope shows the AC input to the modulator. The lower trace shows the frequency-modulated output.

1. Set the amplitude of the function generator to 5 V. Adjust the frequency to produce two full cycles on the upper trace of the oscilloscope display (about 12 kHz).

2. Set the amplitude to its minimum level (1 V). Then step the input voltage upward 1 V at a time by clicking the right arrow on the amplitude control. Notice how the level of the input signal increases. As the input signal level approaches 10 V, notice how the time between peaks on the output signal varies with the input sine wave.

3. Make certain that the frequency of the function generator is still set at about 12 kHz. Set the amplitude to the maximum of 10 V. Sketch the upper and lower traces of the oscilloscope display, using the graph in Figure 25-1 on the Results Sheet.

4. Set the frequency of the function generator to its maximum of 40 kHz. Set the amplitude to the minimum of 1 V.

5. Step the input voltage upward 1 V at a time by clicking the right arrow on the amplitude control. Notice how the time between peaks on the output signal varies with the input sine wave.

PROJECT
25

RESULTS SHEET

PART I   DC INPUT

Table 25-1

dc Voltage In	Frequency Out	Voltage Out
−10		
−8		
−6		
−4		
−2		
0		
2		
4		
6		
8		
10		

STEP 3

Minimum frequency = _____

Maximum frequency = _____

Frequency at ON input = _____

Maximum frequency deviation = _____

STEP 4

Frequency-volt = _____

Questions

1. How does the amount of DC input voltage affect the peak-to-peak output voltage of this circuit?

2. Suppose that this circuit is altered so that the output is 1500 kHz when V_{in} = 0. Assuming that the frequency-volt figure remains unchanged, what will be the minimum and maximum frequencies for this circuit? The maximum frequency deviation?

PART 2   AUDIO INPUT

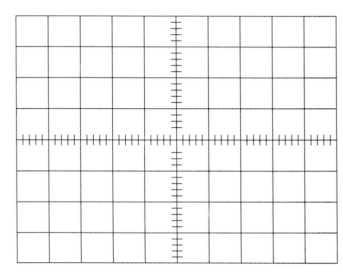

Figure 25-1

Questions

1. How do changes in the amplitude of the modulating signal affect the amplitude of the output signal?

2. In what way does changing the frequency of the modulating signal affect the amplitude of the output signal?

Critical Thinking for Project 25

1. In the circuit of Part 1, the output frequency increases as the input DC level rises in a positive direction. Suppose, however, you want a circuit that shows an increasing output frequency as the DC input level goes more negative. Describe where you would place an inverting DC amplifier in the block diagram in order to make the circuit operate that way.

2. The oscilloscope display for Part 2 of this project shows the FM waveform when the modulating signal is a sine wave. Sketch the same set of waveforms when the modulating signal is a rectangular waveform.

PROJECT 26

FREQUENCY MODULATION

A Hands-On Project

This project uses a voltage-controlled oscillator (VCO) as a frequency modulator. You will:

- Construct the circuit.
- Vary the DC input voltage level and record the corresponding output frequency.
- Plot a graph showing output frequency as a function of input voltage level.

Components and Supplies

2   Resistors, 1 kΩ
1   Resistor, 2.2 kΩ
1   Potentiometer, 10 kΩ
1   Capacitor, 1 nF
1   IC, NC566 voltage-controlled oscillator

Preparation

1. Read Frenzel, *Principles of Electronic Communication Systems*, Section 6-1.

2. Complete the work for Prep Project 25.

Equipment

1   DC power supply
1   Function generator
1   Dual-trace oscilloscope
1   Digital voltmeter (optional)
1   Frequency counter (optional)

LAB PROCEDURE

1. Construct the voltage-to-frequency converter circuit in Figure 26-1. Connect the oscilloscope to monitor the DC input at V_{in} (pin 5), and the signal output at V_{out} (pin 4). Connect other equipment as shown in Figure 26-2.

2. Adjust the potentiometer for each of the DC voltage levels shown in Table 26-1 on the Results Sheet. Record the corresponding V_{in} voltage and V_{out} frequency in the table.

3. Plot the data of Table 26-1, showing the output frequency as a function of input voltage level. Use the graph in Figure 26-3 on the Results Sheet for this plot.

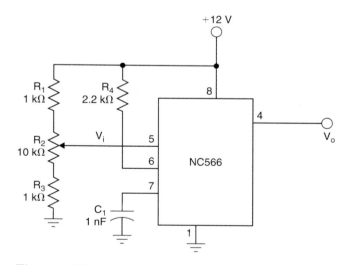

Figure 26-1

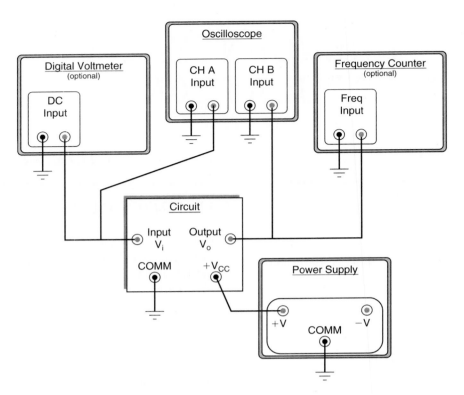

Figure 26-2

PROJECT
26

RESULTS SHEET

Questions

1. From the data in Table 26-1, what is the sensitivity of this modulator in terms of kilohertz per volt?

2. Does the output frequency increase or decrease as the modulating voltage goes more positive?

Table 26-1

DC Voltage In	Frequency Out
2	
4	
6	
8	
10	

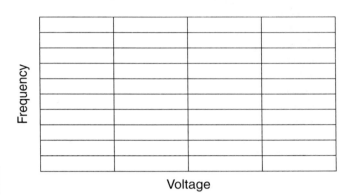

Voltage

Frequency

Figure 26-3

Critical Thinking for Project 26

1. Name the two passive components in Figure 26-1 that determine the center operating frequency of the oscillator.

2. Describe the effect that increasing the value of C_1 from 1 nF to 10 nF will have upon the operation of the circuit.

3. Explain how you would use the 566 VCO to make a sweep-frequency function generator.

Experimental Notes and Calculations

ANALOG FREQUENCY MULTIPLIER

A multiSIM Project

This project simulates the operation of a classic RF frequency multiplier. You will:

- Observe the operation of a basic RF frequency multiplier.
- Alter component values to change the amount of frequency multiplication.

Preparation

Read Frenzel, *Principles of Electronic Communication Systems*, Section 6-1.

Setup Procedure

1. Start multiSIM on your computer.

2. Make sure that your communication lab CD-ROM is in the computer's CD drive.

3. Open the **multiSIM** directory on the CD-ROM.

4. Select **Project_27.msm.**

5. See a multiSIM diagram that is similar to the one shown in Figure 27-1.

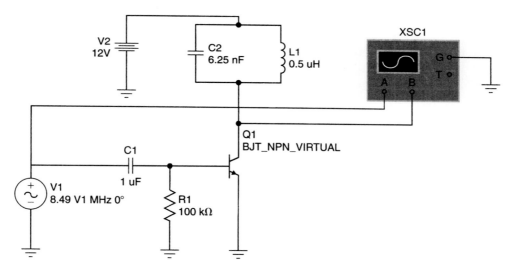

Figure 27-1

This multiplier circuit consists of a class-C amplifier that uses a parallel resonant circuit (or *tank circuit*) as a load. Each time the AC voltage source, V1, "pings" the amplifier, the tank circuit oscillates at its own resonant frequency. If the resonant frequency of the tank circuit happens to be twice the frequency of the AC voltage source, the circuit effectively doubles the source frequency. Likewise, if the tank circuit is tuned to six times the source frequency, the circuit effectively multiplies the input frequency by a factor of six.

LAB PROCEDURE

1. Expand the oscilloscope instrument and start the simulation.

2. Use the oscilloscope to determine the peak-to-peak voltage levels and frequencies of the input and output waveforms when using the default values for V1, C2.

3. Complete the first line of Table 27-1 by providing:
   - Your calculated value for the circuit's resonant frequency (f_r).
   - Your measured values for output amplitude and frequency.
   - The amount of frequency multiplication.

4. Turn off the simulation and set the values for V1, C2, and L1 according to each of the remaining lines in Table 27-1. Provide the information required for each of these sets of values.

PROJECT
27

RESULTS SHEET

Table 27-1

Signal Source (V1)		Tank Circuit			Output Signal		Frequency Multiplier
Amplitude (V$_{p-p}$)	Frequency (MHz)	C2	L1	f$_r$	Amplitude (V$_{p-p}$)	Frequency (MHz)	
6	1	6.25 nF	0.5 μH				
6	1	2.1 nF	0.5 μH				
6	1	1.6 nF	0.5 μH				

Questions

1. Are the calculated resonant frequency and the frequency of the output signal closely related? Nearly equal?

2. Suppose C2 and L1 are set at 6.25 nF and 0.5 μH, respectively. Does the circuit still operation as a frequency doubler if the source frequency is reduced from 1 MHz to 500 kHz?

Critical Thinking for Project 27

1. Can the circuit in Figure 27-1 be used for dividing (as well as multiplying) an input signal frequency?

2. What limits the amount of frequency multiplication that is possible with a single amplifier of this type?

Experimental Notes and Calculations

VARACTOR MODULATOR

An Extended Project

This project uses simulated devices, circuits, and instruments to demonstrate the operation of a varactor diode. You will:

- Gather data and plot a curve showing how the amount of reverse voltage applied to a varactor diode affects its capacitance.
- Gather data and plot response curve for a voltage-to-frequency converter circuit that uses a varactor diode.
- Measure the amount of phase shift caused by changing the amount of reverse bias on a varactor diode operating in a phase modulation circuit.

Preparation

Read Frenzel, *Principles of Electronic Communication Systems*, Section 6-1.

Setup Procedure

1. Select **Extended Projects** from the **Projects** menu.

2. Select **Project 25   Varactor Modulator**.

LAB PROCEDURE

PART 1   DC CAPACITANCE CONTROL

In this part of the project, you will apply a varying amount of DC voltage to a varactor diode and directly read the capacitance. The voltage is supplied by an adjustable DC voltage source. The corresponding capacitance of the varactor diode is monitored by a simulated capacitance meter.

1. Set the DC voltage source to the values listed in Table 28-1 on the Results Sheet. Record the corresponding capacitance values for the varactor diode.

2. Plot the data of Table 28-1 on the graph provided in Figure 28-1.

PART 2   VCO OPERATION

In this part of the project, the varactor diode is connected as the capacitive part of an *LC* oscillator. A DC voltage applied to the varactor diode causes its capacitance to change and, therefore, causes the frequency of the oscillator to change as well.

1. Set the DC voltage source to the values listed in Table 28-2 on the Results Sheet. Record the corresponding capacitance values for the varactor diode.

2. Plot the data of Table 28-2 on the graph provided as Figure 28-2.

PART 3   PM OPERATION

In this part of the project, you will apply a DC voltage to a phase modulation circuit. You can view a schematic diagram of the circuit by clicking the schematic button, located on the right-hand side of the button bar. Also click the schematic button when you want to remove the schematic from the work area.

Refer to the oscilloscope display. The upper waveform shows the applied voltage of the generator. This is the reference waveform. The lower waveform shows the output waveform from the phase-shift circuit—the PM signal. This waveform is shifted with respect to the upper waveform. Moving the mouse pointer across the oscilloscope screen causes a digital display to indicate the phase angle of the reference waveform. Use this feature to determine the phase shift of the PM signal.

1. Move the control on the DC voltage source between its extreme settings. Notice how the phase of the lower waveform is shifted with respect to the upper waveform.

2. Set the DC voltage source control to the values shown in Table 28-3 on the Results Sheet. Estimate the phase angle of the shifted waveform, relative to the reference waveform. Record your results on the same table.

3. Plot a curve on Figure 28-3, showing how the DC input voltage level affects the amount of phase shift.

PART 4 PM OPERATION WITH INVERSION

1. Move the control on the DC voltage source between its extreme settings. Notice how the phase of the lower waveform is shifted with respect to the upper waveform.

2. Set the DC voltage source control to the values shown in Table 28-4 on the Results Sheet. Estimate the phase angle of the shifted waveform, relative to the reference waveform. Record your results on the same table.

3. Plot a curve on Figure 28-4, showing how the DC input voltage level affects the amount of phase shift.

RESULTS SHEET

PART 1   DC CAPACITANCE CONTROL

Table 28-1

Voltage In (V)	Capacitance (pF)
−2	
−4	
−6	
−8	
−10	
−12	
−14	
−16	
−18	
−20	

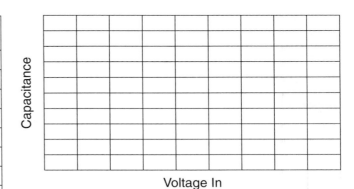

Figure 28-1

Questions

1. From the graph in Figure 28-1, what voltage must be applied to produce a capacitance of 200 pF?

2. According to the data in Figure 28-1, what is the picofarad per volt rating of this varactor diode in the −2-V to −4-V range?

PART 2   VCO OPERATION

Table 28-2

Voltage In (V)	Capacitance (pF)
−2	
−4	
−6	
−8	
−10	
−12	
−14	
−16	
−18	
−20	

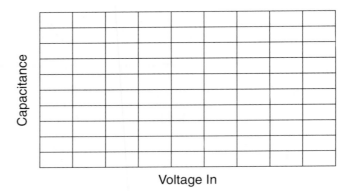

Figure 28-2

Questions

1. From the graph in Figure 28-2, what voltage must be applied to produce a frequency of 15.5 MHz?

2. According to the data of Figure 28-2, what is the approximate megahertz per volt rating of this VCO?

PART 3   PM OPERATION

Note: Use a minus sign (−) to indicate a lagging phase angle and a plus sign (+) to indicate a leading phase angle.

Table 28-3

Voltage In (V)	Phase Angle* (deg)
−2	
−4	
−6	
−8	
−10	
−12	
−14	
−16	
−18	
−20	

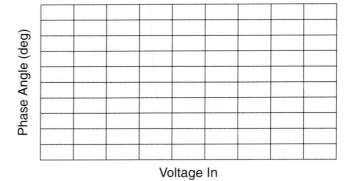

Figure 28-3

Questions

1. Does the lower waveform lead or lag the upper waveform?

2. What DC voltage is required for a phase shift of 57°?

3. What is the conversion (degree per volt) rating of this circuit in the range of −8 V to −20 V?

PART 4 PM OPERATION WITH INVERSION

Note: Use a minus sign (−) to indicate a lagging phase angle, and a plus sign (+) to indicate a leading phase angle.

Table 28-4

Voltage In (V)	Phase Angle* (deg)
−2	
−4	
−6	
−8	
−10	
−12	
−14	
−16	
−18	
−20	

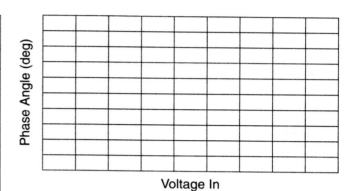

Figure 28-4

Questions

1. Does the lower waveform lead or lag the upper waveform?

2. What DC voltage is required for a phase shift of 64°?

3. What is the conversion (degree per volt) rating of this circuit in the input range of −2 V to −14 V?

Critical Thinking for Project 28

1. Explain the operation of a varactor diode in terms of the width of its depletion layer.

2. Describe the essential differences in the theory of operation between ceramic filters and varactor diodes.

3. This project assumes that the varactor is being operated in its linear region. Explain what this assumption means.

Experimental Notes and Calculations

PROJECT 29

FREQUENCY DEMODULATION

A Prep Project

This project is a computer simulation of tests on a frequency-to-voltage converter. You will:

- Apply various frequency levels to the input of the circuit, and observe the resulting output voltage levels.
- Demonstrate the effect that the input amplitude has on the output voltage levels.
- Plot graphs based on the data gathered from the circuit.

Preparation

Read Frenzel, *Principles of Electronic Communication Systems*, Section 6-3.

Setup Procedure

1. Select **Prep Projects** from the **Projects** menu.

2. Select **Project 29   Frequency Demodulation.**

LAB PROCEDURE

This project uses a function generator to supply various frequencies and voltage levels to a frequency-to-voltage converter circuit. A DC voltmeter monitors the output of the circuit.

1. Set the amplitude of the function generator to 5 V. Adjust the frequency of the function generator to each of the levels shown in Table 29-1 on the Results Sheet. Record the corresponding output voltage levels.

2. Plot a graph showing how the output voltage changes with the input frequency. Use the graph shown in Figure 29-1.

3. With the amplitude of the function generator still set at 5 V, adjust the frequency to obtain an output of 0.0 V. Record this frequency on the Results Sheet.

4. Set the amplitude of the function generator to each of the levels shown in Table 29-2. Record the corresponding output voltage.

5. Plot a graph that shows how the output voltage changes with the input voltage. Use the space provided in Figure 29-2.

PROJECT
29

RESULTS SHEET

Table 29-1

Frequency (MHz)	Voltage Out (V)
400	
410	
420	
430	
440	
450	
460	
470	
480	
490	
500	

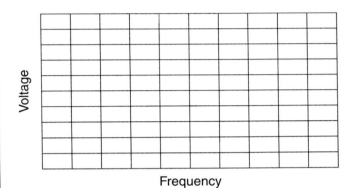

Figure 29-1

STEP 3

Frequency input for 0-V output = _____

Table 29-2

Frequency (MHz)	Voltage Out (V)
1	
2	
3	
4	
5	
6	
7	
8	
9	
10	

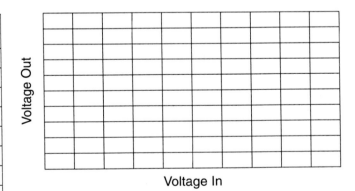

Figure 29-2

Questions

1. From the data in Table 29-2, what is the average volts per megahertz conversion?

2. From the results in Figure 29-2, what effect does a change in input voltage have upon the output voltage?

Critical Thinking for Project 29

1. Explain how the circuit for this project can be used as an FM detector.

2. Suppose that the frequency applied to this circuit is increasing at a steady rate. Describe how the output voltage responds.

3. Suppose that the frequency applied to this circuit drops to 0 Hz. Describe the effect you would most likely notice at the DC output of the circuit.

FREQUENCY DEMODULATION

A Hands-On Project

This project uses a phase-locked loop (PLL) device to demodulate an FM signal generated by the simple modulator used in Project 29. You will:

- Construct the PLL demodulator circuit.
- Determine the VCO frequency of the PLL.
- Plot DC output as a function of input frequency.
- Observe the FM demodulation of a sweep-frequency signal.

Preparation

1. Read Frenzel, *Principles of Electronic Communication Systems*, Section 6-1.

2. Complete the work for Prep Project 29.

Components and Supplies

1	Resistor, 4.7 kΩ
1	Resistor, 47 kΩ
1	Capacitor, 1 nF
1	Capacitor, 10 nF
1	Capacitor, 100 nF
1	IC, LM565 phase-locked loop

Equipment

1	Dual-voltage DC power supply
1	Function generator with a sweep-frequency mode
1	Dual-trace oscilloscope

LAB PROCEDURE

1. Construct the FM demodulator circuit shown in Figure 30-1. Do not at this time connect the function generator to input V_{in}.

2. Calculate the frequency of the VCO from the component values shown in Figure 30-1. Measure the actual VCO frequency at f_{out}. Record both the calculated and the measured frequencies on the Results Sheet.

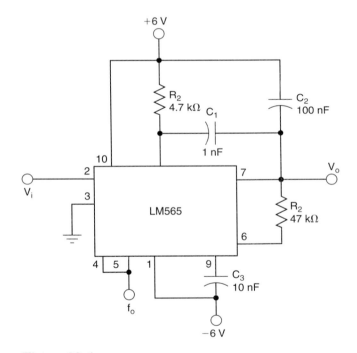

Figure 30-1

3. Adjust the function generator for a sinusoidal output. Set the frequency to the measured value of f_{out} from Step 2, and set the amplitude to 2 V_{p-p}. Apply this signal to input V_{in} of the FM demodulator circuit. Connect the oscilloscope to monitor V_{in} on one channel and V_{out} on the other. Record the frequency at V_{in} and the DC output voltage at V_{out}.

4. Decrease the frequency at V_{in} by 10 percent from the frequency used in Step 3. Record this frequency and the DC output voltage at V_{out}.

5. Increase the frequency at V_{in} by 10 percent from the frequency used in Step 3. Record this frequency and the DC output voltage at V_{out}.

6. Plot the data (V_{out} as a function of f_{in}) from Steps 4 and 5 on the graph provided in Figure 30-2.

7. Switch the operating mode of the function generator to produce a sweep-frequency output. Set the generator's sweep rate to the highest available frequency, and the sweep range to maximum. Observe the output at V_{out} and trigger the oscilloscope from the same signal. Make any necessary slight adjustments in the function generator's sweep range in order to produce the cleanest triangular waveform on the oscilloscope display. This is the FM-demodulated version of the sweep-frequency signal at V_{in}. Sketch the waveform in Figure 30-3.

Name _____ Date _____

RESULTS SHEET

STEP 2

Calculated f_{out} _____

Measured f_{out} = _____

STEP 3

f_{out} _____ V_{out} _____

STEP 4

f_{out} _____ V_{out} _____

STEP 6

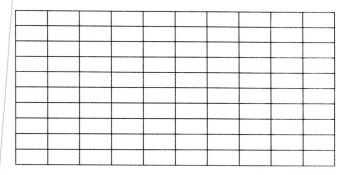

Figure 30-2

STEP 7

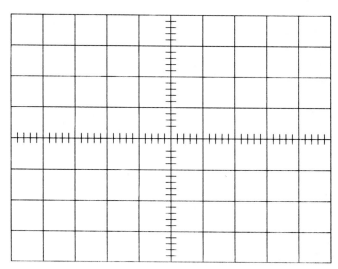

Figure 30-3

Questions

1. What is the formula for estimating the frequency of the VCO in the PLL device when there is no input signal applied?

2. Is the DC output at V_{out} proportional or inversely proportional to the input frequency?

Critical Thinking for Project 30

1. Describe how the circuit arrangement in Step 7 is doing the job of an FM demodulator.

2. Suppose that the sweep range of the input signal in Step 7 exceeds the lower lock range of the PLL circuit. Describe where the resulting distortion will appear on the output waveform at V_{out}.

143

Experimental Notes and Calculations

PROJECT 31

PLL OPERATION

A Prep Project

This project simulates the operation of a phase-locked loop (PLL). An RF generator, a frequency counter, and a digital voltmeter (DVM) provide all the instrumentation required for studying this PLL circuit. You will:

- Determine the free-running frequency of a PLL.
- Determine the lock range of the PLL circuit.
- Determine the capture range of the PLL circuit.
- Observe how changing the input frequency affects the VCO frequency and error voltage level.

Preparation

Read Frenzel, *Principles of Electronic Communication Systems*, Section 6-3.

Setup Procedure

1. Select **Prep Projects** from the **Projects** menu.

2. Select **Project 31   PLL Operation.**

LAB PROCEDURE

When you start this simulation, notice the black rectangle located on the circuit icon. This represents an LED (light-emitting diode) that is sensitive to the circuit's capture mode of operation. This lamp is red when the PLL is in its capture mode. For the purposes of this project, indicate OFF when the lamp is black and indicate ON when the lamp is red.

1. Set the RF generator for its minimum output frequency. Record the readings on all three instruments. Note the status of the LED on the circuit.

2. Click the right arrow button on the frequency adjustment of the generator to increase the frequency at 0.1-MHz steps. Continue this stepping operation until the lamp goes ON (turns to red). At that point, record the status of the instruments.

Note: Do not reduce the frequency setting through Steps 3 and 4. If you happen to overshoot a reading, do not attempt to decrease the frequency in order to recover it. Instead, begin the sequence again at Step 1.

3. Resume clicking the right arrow button on the frequency adjustment of the generator to increase the frequency to 0.1-MHz steps. Continue this stepping operation until the lamp goes OFF (turns to black). At that point, record the status of the instruments.

4. Resume clicking the right arrow button on the frequency adjustment of the generator until you reach the maximum frequency output. Once more, record the status of the instruments.

5. Now begin clicking the left arrow button on the frequency adjustment of the generator to *decrease* the frequency to 0.1-MHz steps. Continue this stepping operation until the lamp goes ON. At that point, record the status of the instruments.

Note: Do not increase the frequency setting through Steps 6 and 7. If you step down beyond a point where you are supposed to stop for a set of readings, restart the operation from Steps 4 and 5.

6. Resume clicking the left arrow button on the frequency adjustment of the generator to decrease the frequency to 0.1-MHz steps, until the lamp goes OFF. At that point, record the status of the instruments.

7. Continue clicking the left arrow button on the frequency adjustment of the generator until you reach the minimum frequency output once again. Record the final status of the instruments.

PROJECT 31

RESULTS SHEET

STEP 1

    Frequency generator = _____

    Frequency counter = _____

    Digital voltmeter = _____

    Circuit lamp status (ON or OFF) = _____

STEP 2

    Frequency generator = _____

    Frequency counter = _____

    Digital voltmeter = _____

    Circuit lamp status (ON or OFF) = _____

STEP 3

    Frequency generator = _____

    Frequency counter = _____

    Digital voltmeter = _____

    Circuit lamp status (ON or OFF) = _____

STEP 4

    Frequency generator = _____

    Frequency counter = _____

    Digital voltmeter = _____

    Circuit lamp status (ON or OFF) = _____

STEP 5

    Frequency generator = _____

    Frequency counter = _____

    Digital voltmeter = _____

    Circuit lamp status (ON or OFF) = _____

STEP 6

    Frequency generator = _____

    Frequency counter = _____

    Digital voltmeter = _____

    Circuit lamp status (ON or OFF) = _____

STEP 7

    Frequency generator = _____

    Frequency counter = _____

    Digital voltmeter = _____

    Circuit lamp status (ON or OFF) = _____

Questions

1. What is the free-running frequency of this PLL?

2. What are the frequencies at the lower and upper limits of the lock range?

3. What are the frequencies at the lower and upper limits of the capture range?

Critical Thinking for Project 31

1. Explain why the LED turned ON at one frequency (when you were stepping the frequency upward) but went OFF at a different frequency (when you were stepping the frequency downward).

2. Explain why the LED turned OFF at one frequency (when you were stepping the frequency upward) but went ON at a different frequency (when you were stepping the frequency downward).

3. Explain why an oscilloscope would be a better instrument for monitoring the error output voltage if you were using a real-time FM signal at the input of the circuit.

PLL OPERATION

A Hands-On Project

The purpose of this project is to investigate the several modes of operation of a phase-locked-loop (PLL) circuit. You will:

• Construct the circuit.
• Determine the VCO frequency.
• Evaluate the capture-lock frequency hysteresis.

Preparation

1. Read Frenzel, *Principles of Electronic Communication Systems*, Section 6-3.

2. Complete the work for Prep Project 31.

Components and Supplies

1   Resistor, 4.7 kΩ
1   Resistor, 47 kΩ
1   Capacitor, 1 nF
1   Capacitor, 10 nF
1   Capacitor, 100 nF
1   IC, LM565 PLL

Equipment

1   Dual-voltage DC power supply
1   Function generator
1   Dual-trace oscilloscope
1   Frequency counter (optional)

Note: The frequency counter is optional because the frequencies for this project can be determined (although somewhat less conveniently) with the oscilloscope.

LAB PROCEDURE

1. Construct the circuit shown in Figure 32-1. Apply power and measure the frequency of the signal at output f_{out}. Record the result on the Results Sheet.

2. Connect the oscilloscope to monitor the frequency at V_{in} on one trace and f_{out} on the other. Set the function generator for:

   Rectangular waveform
   2-V_{p-p} amplitude
   Frequency equal to the frequency of Step 1.

   Connect the function generator to input V_{in}. Trigger the oscilloscope from the signal at V_{in}, and adjust the display to view both waveforms clearly. Sketch the waveform in Figure 32-2.

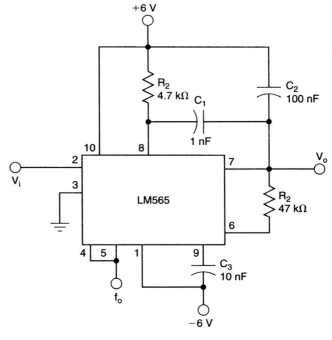

Figure 32-1

3. Gradually increase the frequency of the function generator, noting how the phase changes between the two waveforms. At some point, the PLL will lose its lock on the input waveform—the signal from f_{out} will suddenly appear unstable on the oscilloscope. Record at V_{in} the frequency at which this happens.

4. Gradually decrease the frequency of the function generator, noting the point at which the two waveforms lock once again. Record at V_{in} the frequency at which this relocking occurs.

5. Continue decreasing the frequency of the function generator until lock is lost again. Record the input frequency.

6. Gradually increase the frequency of the function generator until the lock is restored. Record input frequency at this point.

PROJECT
32

RESULTS SHEET

STEP 2

Frequency at V_{in} = _____

STEP 3

Frequency at V_{in} = _____

STEP 4

Frequency at V_{in} = _____

STEP 5

Frequency at V_{in} = _____

STEP 6

Frequency at V_{in} = _____

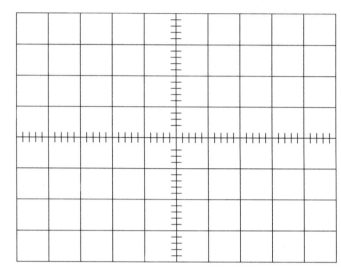

Figure 32-2

Questions

1. According to the data from Steps 3 through 6, what are the upper and lower lock frequencies of this circuit?

2. According to the data from Steps 3 through 6, what are the upper and lower capture frequencies of this circuit?

Critical Thinking for Project 32

1. Explain why it is important to keep the input frequency within the locking range of a PLL when the circuit is being used as an FM demodulator.

2. Describe the effect that a change of ± 20 percent in the amplitude at V_{in} would have upon the signal at V_{out}.

Experimental Notes and Calculations

PROJECT 33 — CRYSTAL OSCILLATOR

A Prep Project

This project is a computer simulation of tests on an NET Pierce crystal oscillator and a TTL crystal oscillator. You will:

- Determine the operating frequency, given the values of the circuit components.
- Determine the actual operating frequency from an oscilloscope display.
- Observe phase relationships in the TTL version.

Preparation

Read Frenzel, *Principles of Electronic Communication Systems*, Section 7-2.

Setup Procedure

1. Select **Prep Projects** from the **Projects** menu.

2. Select **Project 33   Crystal Oscillator.**

LAB PROCEDURE

This project uses interactive schematic diagrams. Moving the mouse pointer to one of the test points on the diagram simulates touching that point with an oscilloscope probe. The waveform for that point appears on the oscilloscope display. Also on the display you will find digital readouts for the frequency and amplitude of the waveform.

PART 1   PIERCE OSCILLATOR

1. Determine the voltage and frequency at all four test points. Record your findings on the Results Sheet.

2. Determine the phase differences between the signals at TP 1 and TP 2, TP 2 and TP 3, and TP 3 and TP 4. Record your data on the results sheet.

PART 2   TTL CRYSTAL OSCILLATOR

1. Determine the voltage and frequency at all four test points. Record your findings on the Results Sheet.

2. Determine the phase differences between the signals at TP 1 and TP 2, TP 2 and TP 3, and TP 3 and TP 4. Record your data on the Results Sheet.

Experimental Notes and Calculations

Name _____ Date _____

PROJECT
33

RESULTS SHEET

PART 1   PIERCE OSCILLATOR

STEP 1

TP 1 voltage = _____

TP 1 frequency = _____

TP 2 voltage = _____

TP 2 frequency = _____

TP 3 voltage = _____

TP 3 frequency = _____

STEP 2

Phase difference between TP 1 and TP 2 =

_____

Phase difference between TP 1 and TP 2 =

_____

Questions

1. What is the phase relationship between the signals at TP 1 and TP 2?

2. What is the phase relationship between the signals at TP 2 and TP 3?

PART 2   TTL CRYSTAL OSCILLATOR

STEP 1

TP 1 voltage = _____

TP 1 frequency = _____

TP 2 voltage = _____

TP 2 frequency = _____

TP 3 voltage = _____

TP 3 frequency = _____

STEP 2

Phase difference between TP 1 and TP 2 =

_____

Phase difference between TP 1 and TP 2 =

_____

Questions

1. Which components on the schematic diagram determine the operating frequency of this circuit?

2. What is the voltage gain of the circuit between points TP 1 and TP 2? Between TP 3 and TP 4?

Critical Thinking for Project 33

1. Describe the operation of a crystal oscillator in a circuit in which the amplifier stage has a voltage gain of less than 1.

2. Explain how IC1-A, IC1-B, and XTAL in the circuit for Part 2 make up a positive feedback loop.

3. Describe the operation of the circuit for Part 2 if the feedback loop included IC1-C as well as IC1-A, IC1-B, and the crystal.

CRYSTAL OSCILLATOR

A Hands-On Project

The circuits in this project use an NET as the active element of a Pierce oscillator, and TTL logic inverters as the active elements in a simple crystal oscillator. You will:

- Construct the circuits.
- Measure operating frequencies.
- Gather data for determining the oscillators' frequency stability.

Preparation

1. Read Frenzel, *Principles of Electronic Communication Systems*, Section 7-2.

2. Complete the work for Prep Project 33.

Components and Supplies

2   Resistors, 330 Ω
1   Resistor, 100 kΩ
1   Capacitor, 10 nF
1   Capacitor, 100 nF
2   Capacitors, 100 pF
1   JFET, 2N5457
1   IC, 7404 TTL hex inverter
1   Crystal, 3.579 MHz
1   Crystal, 5.00 MHz

Equipment

1   Variable DC power supply
1   Oscilloscope
1   Frequency counter (optional)

LAB PROCEDURE

PART I   PIERCE CRYSTAL OSCILLATOR

1. Construct the Pierce oscillator circuit shown in Figure 34-1. Adjust the supply voltage for +12 V, and then measure and record the frequency and signal level at V_{out}.

2. For each of the entries in Table 34-1, set the power supply to the given voltage level and record the frequency and peak-to-peak value of the waveform at V_{out}.

3. Plot the frequency data of Step 2 on the graph in Figure 34-3.

4. Replace the 3.578-MHz crystal in the circuit with a 5.00-MHz crystal. Measure and record the frequency and signal level at V_{out}.

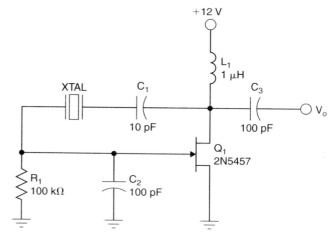

Figure 34-1

PART 2   TTL CRYSTAL OSCILLATOR

1. Construct the TTL oscillator circuit shown in Figure 34-2. Make sure that you use the +5-Vdc terminals on the power supply, and apply power to the circuit. Connect the oscilloscope to output V_{out}, and then measure and record the frequency and signal level at V_{out}.

2. Observe the waveforms at the outputs of IC1-A, IC1-B, and IC1-C. Sketch these three waveforms in the spaces provided in Figure 34-4, making sure that your drawings show the phase relationships among these three points in the circuit.

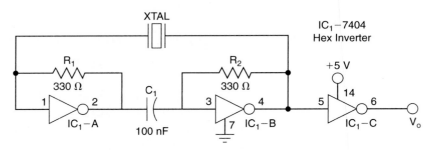

Figure 34-2

PROJECT 34

RESULTS SHEET

PART I PIERCE CRYSTAL OSCILLATOR

STEP 1

f_{out} = _____ V_{out} = _____ V_{p-p}

STEP 4

f_{out} = _____ V_{out} = _____ V_{p-p}

Questions

1. Based on the data in Figure 34-3, can you say that a Pierce crystal oscillator is frequency-stable with regard to changes in the power supply voltage? Explain your answer.

2. Assuming that the 5.00-MHz crystal in this project has a rated precision of ±50 ppm, what are the minimum and maximum allowable frequencies?

Table 34-1

dc Supply Voltage	Output Voltage (V_{p-p})	Output Frequency
+2 V		
+4 V		
+6 V		
+8 V		
+10 V		
+12 V		

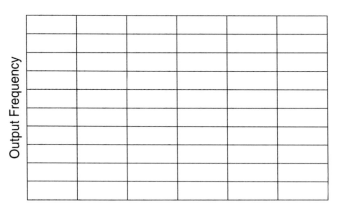

Figure 34-3

PART 2 TTL CRYSTAL OSCILLATOR

STEP 1

f_{out} = _____ V_{out} = _____

Questions

1. Which inverters are included in the positive feedback loop for this circuit?

2. Which component in this circuit has the greatest amount of influence on the operating frequency?

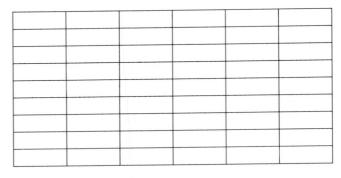

Figure 34-4

Critical Thinking for Project 34

1. One of the advantages of crystal oscillators is that their operating frequency can be changed by simply changing the crystal. Explain what must be done to an *LC* oscillator, such as a Colpitts or Hartley oscillator, to change its operating frequency.

2. Explain why it is important that frequency sources in communication circuits are stable with respect to changes in power supply voltage and output loading.

TUNED AMPLIFIERS

A Prep Project

This project is a computer simulation of a class C tuned amplifier. You will:

- Determine the output frequency and voltage gain of a single-stage tuned amplifier.
- Determine the frequency response characteristics of the amplifier.
- Observe the bandpass characteristics of a two-stage tuned amplifier.

Preparation

Read Frenzel, *Principles of Electronic Communication Systems*, Section 7-3.

Setup Procedure

1. Select **Prep Projects** from the Projects.

2. Select **Project 35   Tuned Amplifiers**.

LAB PROCEDURE

In both parts of this project, the RF generator is connected to the input of the circuit and to the upper trace of the oscilloscope. This connection provides the input signal as well as a reference signal for the display. The lower trace of the oscilloscope is connected to a simulated probe.

The project uses interactive block diagrams. You can select a test point to be monitored on the lower trace of the oscilloscope by clicking the point with the mouse. The probe is attached to the test point that has the dark blue label.

PART 1   SINGLE-STAGE RF AMPLIFIER

For this part of the project, the RF generator and schematic diagram compete for viewing space on the workbench. Clicking the main body of either one brings it to the foreground for better viewing.

1. Calculate the center frequency f_c of the tuned circuit. Record your result on the Results Sheet.

2. Attach the probe to TP 1 of the schematic diagram. Adjust the amplitude of the RF generator to 24 V. Set the RF generator to your calculated center frequency of the collector circuit. Assuming that the vertical sensitivity of the oscilloscope is 24 V/div, record the amplitudes of the RF generator output (channel A) and the signal at TP 1 (channel B).

3. Attach the probe to the output of the amplifier at TP 2. Adjust the frequency of the RF generator slightly to make sure you are at or very close to the center frequency.

4. Working from the peak voltage level at TP 2, calculate the upper and lower cutoff voltage levels. Record these levels on the Results Sheet.

5. While monitoring the output at TP 2, adjust the input frequency to the cutoff voltage levels and record the corresponding frequencies on the Results Sheet.

PART 2   TWO-STAGE RF AMPLIFIER

1. Fix the probe to TP 1 on the interactive diagram. Adjust the amplitude of the RF generator to 24.0 V.

2. Move the probe to TP 2, and adjust the frequency of the RF generator to obtain a peak voltage level from tuned amplifier 1. Record the frequency from the RF generator and the voltage appearing at TP 2. (Assume that the oscilloscope is scaled at 24 V/div.)

3. Determine, by measurement, the upper and lower cutoff frequencies of tuned amplifier 1. Record your findings on the Results Sheet.

4. Move the probe to the output of tuned amplifier 2 (TP 3). Determine the center frequency and the upper and lower cutoff frequencies for the output at TP 3. Record your data on the Results Sheet.

PROJECT
35

Name _____ Date _____

RESULTS SHEET

PART 1 SINGLE-STAGE RF AMPLIFIER

f_c = _____

STEP 2

RF generator voltage = _____

TP 1 voltage = _____

STEP 3

RF generator voltage = _____

TP 2 voltage = _____

STEP 4

Calculated lower cutoff voltage = _____

Calculated upper cutoff voltage = _____

STEP 5

Measured lower cutoff frequency = _____

Measured upper cutoff frequency = _____

Questions

1. Why are the two waveforms of Step 1 identical?

2. How do you know in Step 3 that the input is set near or at the center frequency of the amplifier?

3. What is the bandwidth of the circuit?

PART 2 TWO-STAGE RF AMPLIFIER

STEP 2

Center frequency of tuned amplifier 1 =

_____

Peak voltage of tuned amplifier 1 = _____

STEP 3

Lower cutoff frequency of tuned amplifier 1 =

_____

Upper cutoff frequency of tuned amplifier 1 =

_____

STEP 4

Center frequency at TP 3 = _____

Lower cutoff frequency at TP 3 = _____

Upper cutoff frequency at TP 3 = _____

Questions

1. How did you determine the upper and lower cutoff frequencies of tuned amplifier 1 in Step 3?

2. What is the bandwidth of the signal at TP 2? At TP 3?

Critical Thinking for Project 35

1. Describe the similarities between the response curves for an active bandpass filter circuit and a class C tuned amplifier.

2. The center frequencies for the amplifiers in a two-stage tuned amplifier might be slightly different. Why would this discrepancy be intentional?

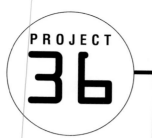

PROJECT 36

TUNED AMPLIFIERS

A Hands-On Project

This project uses a single-stage class C transistor amplifier that has a tuned circuit as its collector load. You will:

- Construct the circuit.
- Determine the output frequency and voltage gain.
- Determine the frequency response characteristics of the amplifier.
- Tune the amplifier for maximum gain.

Preparation

1. Read Frenzel, *Principles of Electronic Communication Systems*, Section 7-3.

2. Complete the work for Prep Project 35.

Components and Supplies

1	2N3904 or 2N2222 Resistor, 100 Ω
1	Resistor, 100 kΩ
1	Capacitor, 10 nF
2	Capacitors, 100 nF
1	Trimmer capacitor, 20–90 pF
1	Inductor, 1 mH
1	NPN transistor

Equipment

1	DC power supply
1	Dual-trace oscilloscope
1	Function generator
1	Frequency counter (optional)

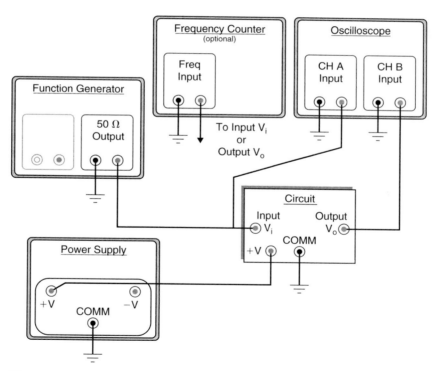

Figure 36-1

PART 1 VARIABLE INPUT FREQUENCY, FIXED TANK FREQUENCY

1. Construct the class C amplifier circuit shown in Figure 36-2. Connect the function generator, oscilloscope, and optional frequency counter as shown in Figure 36-1. Adjust the function generator for a 500-mV$_{p-p}$ sinusoidal waveform.

2. Calculate the resonant frequency of the LC circuit. Record your result on the Results Sheet.

3. For each of the entries in Table 36-1, set the function generator for the given frequency, double-check the value of V_{in} (500-mV$_{p-p}$), and record the peak-to-peak value of V_{out}.

4. Within the range of frequencies in Table 36-1, locate the circuit's actual resonant frequency by adjusting the function generator to obtain the peak output signal level at V_{out}. Record the frequency and output voltage level.

5. Using $V_{out(max)}$ as the 0-dB level, calculate the dB loss for each of the frequencies in Table 36-1. Then plot the circuit's response curve on the two-cycle semilog graph in Figure 36-3.

PART 2 FIXED INPUT FREQUENCY, VARIABLE TANK FREQUENCY

1. Modify the circuit in Figure 36-2 by replacing fixed capacitor C_3 with a trimmer capacitor. Connect the function generator and oscilloscope to the circuit as shown in Part 1.

2. Adjust the function generator for a 700-kHz, 500-mV$_{p-p}$ sinusoidal waveform.

3. Carefully adjust the trimmer capacitor, looking for a peak output waveform. Record the value of $V_{out(max)}$ on the Results Sheet.

4. Adjust the amplitude of the function generator at V_{in} to obtain the largest undistorted signal at V_{out}. Record the resulting values of V_{in} and V_{out}. Calculate and record the voltage gain of the amplifier, based on the voltage values you've just obtained.

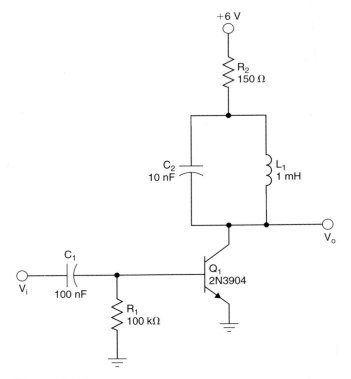

Figure 36-2

RESULTS SHEET

PROJECT 36

PART 1 VARIABLE INPUT FREQUENCY, FIXED TANK FREQUENCY

STEP 2

Calculated f_c = _____

Table 36-1

f (kHz)	v_o (V$_{p-p}$)	v_o/v_i	$20 \log(v_o/v_i)$ (dB)
6			
8			
10			
30			
50			
70			
90			
200			
400			
500			

PART 2 FIXED INPUT FREQUENCY, VARIABLE TANK FREQUENCY

STEP 3

$V_{out(max)}$ = _____

STEP 4

V_{in} = _____ V_{out} = _____

A_v = _____

Questions

1. What is the voltage gain of this amplifier, expressed in dB?

2. What range of center frequencies is available for this circuit, where the inductor is fixed at 1 mH and C_2 is variable between 20 pF and 90 pF?

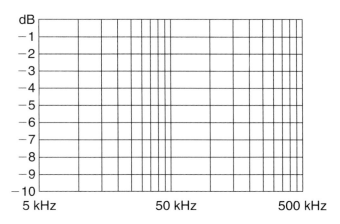

Figure 36-3

Critical Thinking for Project 36

1. Explain how you can tell from the schematic that the amplifier is biased for class B or class C operation.

2. Which components in these circuits are responsible for the flywheel effect?

3. Explain why this circuit is most efficient at the resonant frequency of the tank circuit.

TUNING RF AMPLIFIER CIRCUITS

An Extended Project

This project simulates the behavior of transformer-coupled AC amplifiers. You will:

- Tune the primary and secondary windings of a two-stage, tuned-transformer circuit.
- Tune the windings of a three-stage, tuned-transformer amplifier circuit.

Preparation

Read Frenzel, *Principles of Electronic Communication Systems*, Section 7-3.

Setup Procedure

1. Select **Extended Projects** from the **Projects** menu.

2. Select **Project 37   Tuning RF Amplifier Circuits.**

LAB PROCEDURE

This project uses simulated tunable transformer windings that can be trimmed to produce a peak output at a specified frequency. An RF generator acts as the signal source, and an analog RF voltmeter registers the circuit's output amplitude.

The primary and secondary windings of the transformers are tuned by depressing a mouse key, and the mouse pointer is located on a rectangle labeled CW or CCW. This action simulates the turning of the adjustment screw of a trimmer capacitor or the turning of a threaded powdered iron slug of a variable inductor. Pressing the left mouse key turns the screw in the counterclockwise (CCW) direction. Pressing the right mouse key turns the screw in the clockwise (CW) direction. When you "hit the limit" for the screw in one direction, the label turns red and a warning bell sounds. Further adjustment is possible only in the opposite direction.

PART 1   TWO-STAGE AMPLIFIER

1. Set the RF generator to 42.5 MHz.

2. Adjust the secondary winding (SEC) to peak the output voltage. Record the approximate indication of the meter.

3. Adjust the primary winding (PRI) to peak the output voltage. Record the approximate indication of the meter.

The coupling transformer is now tuned to 42.5 MHz. In the next series of steps, you will determine the bandwidth of the transformer.

4. Using the voltage reading from Step 3 as the maximum output level, calculate the voltage level for the upper and lower half-power points of the circuit's response curve. Record your values on the Results Sheet.

5. Vary the output of the RF generator to determine the upper and lower cutoff frequencies. Record these frequencies on the Results Sheet.

PART 2 THREE-STAGE AMPLIFIER

1. Set the RF generator for 50.2 MHz.

2. Tune transformer T_2 for a peak output voltage. Be sure to tune the secondary winding first. When you have tuned both windings, record the peak output voltage on the Results Sheet.

3. Tune transformer T_1 for a peak output voltage (secondary winding, followed by the primary winding). Record the peak output level on the Results Sheet.

4. Carefully tweak the four windings to make sure that you are tuned to the highest possible output level. Record this peak output level.

5. Use the voltage of Step 4 as the 0-dB output level, and calculate the voltage output for the upper and lower −3-dB frequencies. Record this voltage level on the Results Sheet.

6. Vary the output frequency of the RF generator to determine the actual upper and lower cutoff frequencies. Record these frequencies on the Results Sheet.

RESULTS SHEET

PART 1 TWO-STAGE AMPLIFIER

STEP 2

Relative voltage out = _____

STEP 3

Relative voltage out = _____

STEP 4

Calculated half-power points voltage = _____

STEP 5

Measured lower cutoff frequency = _____

Measured upper cutoff frequency = _____

Questions

1. What is the bandwidth of this circuit?

2. What is the Q of this circuit?

PART 2 THREE-STAGE AMPLIFIER

STEP 2

Relative voltage out = _____

STEP 3

Relative voltage out = _____

STEP 4

Relative voltage out = _____

STEP 5

Calculated half-power points voltage = _____

STEP 6

Measured lower cutoff frequency = _____

Measured upper cutoff frequency = _____

Questions

1. What is the bandwidth of this circuit?

2. What is the Q of this circuit?

Critical Thinking for Project 37

1. Describe how you would broaden the bandwidth of the circuit for Part 1.

2. Prior to the mid-1980s, one common problem with television receivers was routine intermediate frequency (IF) alignment. Explain what this problem meant and why it happened.

FREQUENCY MULTIPLIERS

A Prep Project

This project is a computer simulation of 1- and 2-stage frequency multipliers. You will:

- Note the gain and operating frequencies of a single-stage frequency doubler and tripler.
- Observe the output waveform of a single-stage voltage quadrupler.
- Note the voltage gains, operating frequencies, and waveforms.

Preparation

Read Frenzel, *Principles of Electronic Communication Systems*, Section 7-3.

Setup Procedure

1. Select **Prep Projects** from the **Projects** menu.

2. Select **Project 38   Frequency Multipliers.**

LAB PROCEDURE

This project uses interactive schematic diagrams. A simulated test probe is connected to a frequency counter and a digital AC meter. Connect the probe to a test point by clicking the test point label.

PART 1   TUNED AMPLIFIER AS A FREQUENCY

1. Set the amplitude of the RF generator to 10.0 V. Carefully sweep the frequency of the RF generator from its minimum to its maximum limits, watching the readings on the AC digital voltmeter. Note the location of all peaks in this output voltage. For each peak you find (there are five of them), record the frequency setting of the RF generator (f_{in}), the frequency on the frequency counter (f_{out}), and the voltage on the AC digital voltmeter (V_{out}).

2. Double-check your results by sweeping the frequency from its maximum to its minimum limit. Resolve any differences you find in the peaks from Step 1.

PART 2   SINGLE-STAGE FREQUENCY MULTIPLIER

In this part of the project, the input frequency and amplitude are fixed (as they would be when a frequency multiplier is being used for stepping up the frequency of a crystal oscillator circuit).

1. Connect the test probe to TP 2, and repeatedly click the pushbutton on the filter selector. Note the changes in frequency and amplitude at TP 2.

2. Set the filter selector to the five values shown in Table 38-1. In each case, measure the frequencies and amplitudes at test points TP 1 and TP 2. Also calculate the frequency multiplication factor.

PART 3   TWO-STAGE FREQUENCY MULTIPLIER

1. Note that the input to this circuit is fixed at 3.5 MHz, 4.8 V_{p-p}. Connect the test probe to the TP 1. Record the output frequency and amplitude on the Results Sheet.

2. Move the test probe to test points TP 2 and TP 3. Record the frequencies and amplitudes that you find.

RESULTS SHEET

PART 1 TUNED AMPLIFIER AS A FREQUENCY

STEP 1

First peak f_{in} = _____ f_{out} = _____

V_{out} = _____

Second peak f_{in} = _____ f_{out} = _____

V_{out} = _____

Third peak f_{in} = _____ f_{out} = _____

V_{out} = _____

Fourth peak f_{in} = _____ f_{out} = _____

V_{out} = _____

Fifth peak f_{in} = _____ f_{out} = _____

V_{out} = _____

Questions

1. What is the ratio $f_{in} : f_{out}$ for each peak?

2. What is the dB gain of the circuit for each peak?

PART 2 SINGLE-STAGE FREQUENCY MULTIPLIER

Table 38-1

Filter Selector	TP 1 f_i	TP 2 v_i	TP 3 f_o	TP 1 v_o	Frequency Multiplier
50 MHz					
100 MHz					
150 MHz					
200 MHz					
250 MHz					

Questions

1. What is the dB gain of the circuit for each setting?

2. What is the relationship between the amount of frequency multiplication and the voltage gain of this amplifier?

PART 3   TWO-STAGE FREQUENCY MULTIPLIER

VALUES AT TP 1:

    Frequency = _____

    Amplitude = _____

VALUES AT TP 2:

    Frequency = _____

    Amplitude = _____

VALUES AT TP 3:

    Frequency = _____

    Amplitude = _____

Questions

1. What is the frequency multiplication factor for tuned amplifier 1?

2. What is the voltage gain of tuned amplifier 1?

3. What is the frequency multiplication factor for tuned amplifier 2?

4. What is the voltage gain of tuned amplifier 2?

5. What is the overall frequency multiplication factor of the circuit?

6. What is the overall voltage gain of the circuit?

Critical Thinking for Project 38

1. Explain why the output voltage from a frequency multiplier decreases as the multiplication ratio increases.

2. Resolve the apparent discrepancy between the results you find when working with frequency-multiplier amplifiers and the fact that a pure sinusoidal waveform has no harmonics beyond the fundamental frequency.

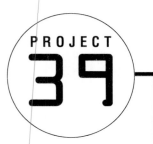

PROJECT 39

FREQUENCY MULTIPLIERS

A Hands-On Project

This project uses a single-stage bipolar transistor amplifier to demonstrate the operation of frequency multipliers. You will:

- Construct the circuit.
- Fine-tune the circuit.
- Apply frequencies and record output frequencies.

Preparation

1. Read Frenzel, *Principles of Electronic Communication Systems*, Section 7-3.

2. Complete the work for Prep Project 38.

Components and Supplies

1	Resistor, 150 Ω
1	Resistor, 100 kΩ
1	Capacitor, 1 μF
1	Capacitor, 100 nF
1	Capacitor, 100 pF
1	Trimmer capacitor, 20–90 pF
1	Inductor, 1 mH
1	NPN transistor, 2N3904 or 2N2222

Equipment

1	Variable power supply
1	Function generator
1	Dual-trace oscilloscope
1	Frequency counter (optional)

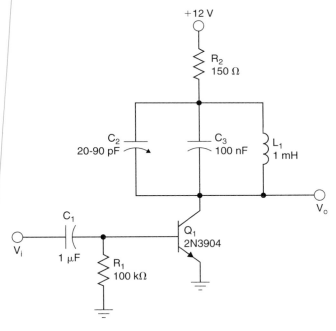

Figure 39-1

LAB PROCEDURE

1. Construct the class C amplifier circuit shown in Figure 39-1. Connect the function generator to V_{in}, and adjust it for a 3-V_{p-p} sinusoidal waveform at 455 kHz. Connect the dual-trace oscilloscope to input V_{in} and to output V_{out}.

2. Adjust the trimmer capacitor to obtain the peak output signal at V_{out}. Record the output frequency f_{out} in the first row of Table 39-1.

   Note: The amplifier should now be tuned for maximum gain at 455 kHz.

3. Adjust the frequency of the function generator for 455 kHz/2, or 227.5 kHz, as indicated in the second row of Table 39-1. Record the output frequency f_{out} and ratio of input to output frequency.

4. Repeat the operations in Step 3 for the remaining rows in Table 39-1.

   Note: A more stable waveform can be obtained by triggering the horizontal sweep on the oscilloscope from the input signal of the circuit.

5. Sketch the input and output waveforms for the 1:10 frequency multiplication ratio in Figure 39-2.

Name _____ Date _____

RESULTS SHEET

Questions

1. Do any of the setups in Table 39-1 represent the operation of a frequency tripler? Explain your answer.

2. What are the minimum and maximum resonant frequencies that can be obtained from the circuit in Figure 39-1, given its values for L_1, C_2, and C_3?

Table 39-1

Input Frequency (f_i)	Output Frequency (f_o)	Frequency Ratio $(f_i : f_o)$
455 kHz		
455/2 kHz		
455/4 kHz		
455/8 kHz		
455/10 kHz		

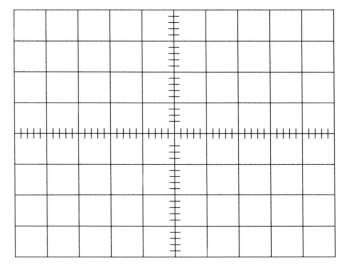

Figure 39-2

Critical Thinking for Project 39

1. Explain the purpose of the tuned circuit in the collector.

2. Determine the overall multiplication factor of a four-stage multiplier where the individual stages have frequency multiplication factors of 2, 3, 4, and 6.

3. Explain why a class C amplifier will always distort a sinusoidal input waveform, and explain why this fact is used to advantage in a frequency multiplier circuit.

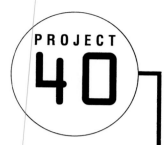

IMPEDANCE-MATCHING NETWORKS

A Prep Project

This project simulates the operation of an impedance matching network. At a desired frequency, the RF signal source has an internal impedance that is less than the load impedance. You will:

- Show that an L network acts as a parallel tuned circuit when it is matching a lower to a higher impedance.
- Determine the intended operating frequency for a particular L network.

Preparation

Read Frenzel, *Principles of Electronic Communication Systems*, Section 7-4.

Setup Procedure

1. Select Prep **Projects** from the **Projects** menu.

2. Select **Project 40   Impedance-Matching Networks.**

LAB PROCEDURE

This project uses an interactive schematic diagram. The AC voltmeter can be connected to one of the test points, TP 1 or TP 2. You make this selection by clicking the desired test point with the mouse pointer. Also, you can open or close the connection between the signal source and the rest of the circuit. Open or close that connection by clicking the load button, located near the bottom of the schematic diagram. You will know that the connection is open when you see a break in the conductor between R_g and L.

1. Select TP 1 and open the connection between the signal source and load. This is the no-load condition for the signal source.

2. Set the frequency of the function generator to 200 MHz, and the amplitude to each of the settings shown in Table 40-1. For each of these settings, check and record the voltages at TP 1 and TP 2.

3. Set the amplitude of the function generator to 10.0 V and select TP 1. Make sure that the load circuit is open. Move the frequency adjustment between its two extremes, and note the effect upon the voltage at TP 1.

4. Click the load button to close the circuit between the signal source and load. Select TP 2 for display on the digital voltmeter, and set the amplitude of the function generator to 10.0 V.

5. Sweep the frequency setting of the function generator until you find a peak reading at TP 2. Record the function generator frequency and voltage reading at TP 2 on the Results Sheet. Select TP 1 and record the voltage found at that point.

6. Make sure that the circuit is still closed between the signal source and load, and that the amplitude of the function generator is still set to 10.0 V. Adjust the frequency of the function generator for a frequency that is well away from the center frequency of Step 5. Record this frequency and the voltages at TP 1 and TP 2.

PROJECT 40

RESULTS SHEET

STEP 3

_____

_____

_____

STEP 5

Center frequency = _____

Voltage at TP 2 = _____

Voltage at TP 1 = _____

STEP 6

Frequency = _____

Voltage at TP 2 = _____

Voltage at TP 1 = _____

Table 40-1

Amplitude Setting	Voltage TP 1	Voltage TP 2
0		
5		
10		
15		
20		
25		
30		

Questions

1. How do you account for the results of Step 3?

2. The input and output impedances match at only one frequency. According to the data from this project, what is that frequency?

3. What is the ratio of voltages at TP 1 and TP 2 according to the data of Step 5?

Critical Thinking for Project 40

1. Explain the results shown in Table 40-1.

2. Describe how you would use this project to determine the bandwidth of the impedance-matching effect.

3. Explain why the voltage at TP 1 and TP 2 are equal only at the resonant frequency of the impedance-matching network.

Experimental Notes and Calculations

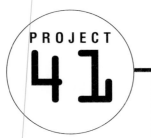

IMPEDANCE-MATCHING NETWORKS

PROJECT 41

A Hands-On Project

The objective of this project is to give you practical experience with *LC* impedance-matching networks. You will:

- Construct the circuit.
- Determine the frequency at which impedances match.
- Determine changes required for operating the circuit at a different frequency.

Preparation

1. Read Frenzel, *Principles of Electronic Communication Systems*, Section 7-4.

2. Complete the work for Prep Project 35.

Components and Supplies

1   Resistor, 300 Ω
1   Inductor, 1 mH
1   Capacitor, 68 nF

Equipment

1   Function generator
1   Dual-trace oscilloscope
1   Frequency counter (optional)

LAB PROCEDURE

The output impedance of a function generator is typically 50 Ω. The objective of this demonstration is to match this 50-Ω output with a 300-Ω load resistance. The type of impedance-matching network used in this project operates to specification only within a range of frequencies. In this example, optimum impedance matching takes place around 17.8 kHz.

1. Construct the external circuit (made up of *L*, *C*, and *RL)* as shown in Figure 41-1. Do not connect this circuit to the function generator at this time, however.

2. Connect the oscilloscope to the output of the function generator and adjust the generator for a sinusoidal waveform at about 17.8 kHz and 10 V_{p-p}. Record the voltage level on the Results Sheet as the unloaded AC output level.

3. Connect the circuit to the function generator to complete the circuit shown in Figure 41-1. Make sure that the oscilloscope is still connected across the output of the function generator.

4. Carefully adjust the fine tuning of the frequency generator to obtain an output signal that is exactly one-half the value of the unloaded output voltage (Step 2). Record this voltage and the frequency at which it occurs.

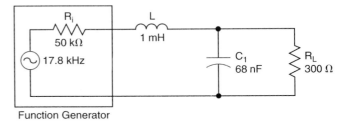

Figure 41-1

PROJECT
41

RESULTS SHEET

STEP 2

Unloaded generator voltage = _____

STEP 4

Output voltage at matched impedance = _____

Frequency of matched impedance = _____

Questions

1. There might be a considerable amount of difference between the design frequency for this circuit and the frequency you find in Step 4. How do you account for this discrepancy?

2. What is the phase angle between the current and voltage at *RL* when this circuit is properly matched?

Critical Thinking for Project 41

1. Explain the advantage of using a low-pass filter (such as the one used in this circuit) for impedance matching, as opposed to a high-pass filter.

2. Explain why source-load impedance matching can be important for:

   a. Test instruments
   b. Transmitters and antennas

D/A CONVERSION

A Prep Project

This project simulates the operation of a 4-bit digital-to-analog (D/A) converter circuit. You will:

- Apply binary values to the input and note the output voltage levels.
- Observe the effect of bias voltage on the output voltage levels.
- Calculate and verify the resolution of the converter.

Preparation

Read Frenzel, *Principles of Electronic Communication Systems*, Section 8-2.

Setup Procedure

1. Select **Prep Projects** from the **Projects** menu.

2. Select **Project 42   D/A Conversion.**

LAB PROCEDURE

The DC voltage source is a variable DC source that has an output range of 0 V to 23.9 V. In this project, the DC voltage source serves as the reference voltage (VR) for the D/A converter. The digital source provides the digital input for the circuit. This is a simple 4-bit binary source. You can toggle each of the binary values by clicking the corresponding data switch.

The DC voltmeter indicates the actual analog output voltage from the D/A circuit.

1. Adjust the DC voltage source for 15.0 V. Calculate the resolution of this circuit, and enter your response on the Results Sheet.

2. Enter each of the binary values cited in Table 42-1. Record the output voltage reading on the DC voltmeter for each of these values.

3. Graph the data of Table 42-1 in the space provided in Figure 42-1.

PROJECT 42

RESULTS SHEET

STEP 1

Calculated resolution = _____

Questions

1. Based on the data in Table 42-1, what is the actual minimum output voltage?

2. What is the actual maximum output voltage?

3. What is the actual resolution of this circuit?

Table 42-1

D_3	D_2	D_1	D_0	v_o
0	0	0	0	
0	0	0	1	
0	0	1	0	
0	0	1	1	
0	1	0	0	
0	1	0	1	
0	1	1	0	
0	1	1	1	
1	0	0	0	
1	0	0	1	
1	0	1	0	
1	0	1	1	
1	1	0	0	
1	1	0	1	
1	1	1	0	
1	1	1	1	

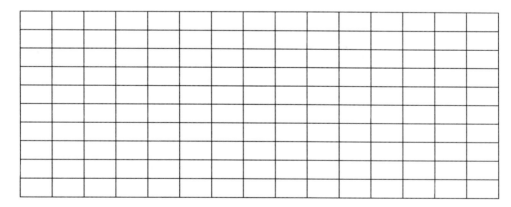

Figure 42-1

Critical Thinking for Project 42

1. Explain why the resolution of an 8-bit D/A converter is inherently greater than the resolution of a 4-bit D/A converter. How much greater is the resolution?

2. Describe what would happen to the resolution of this circuit if the reference voltage were reduced from 15 V to 7.5 V.

D/A CONVERSION

A Hands-On Project

This project demonstrates the operation of a D/A converter circuit using a four-input R-2R circuit. Although this particular circuit configuration is generally considered to be obsolete, there is no simpler and clearer way to demonstrate the principles of D/A conversion. In this project you will:

- Construct the circuits.
- Determine the analog output voltage level as a function of binary input values.
- Observe and explain the analog output voltage when a high-frequency binary counter is applied to the inputs.

Preparation

1. Read Frenzel, *Principles of Electronic Communication Systems*, Section 8-2.

2. Complete the work for Prep Project 42.

Components and Supplies

5	Resistors, 10 kΩ
4	Resistors, 20 kΩ
1	IC, 741 op amp
1	IC, 7493 binary counter
4	SPDT switches

Note: Jumper wires may be used in place of SPDT switches.

Equipment

1	DC power supply
1	Oscilloscope
1	Function generator

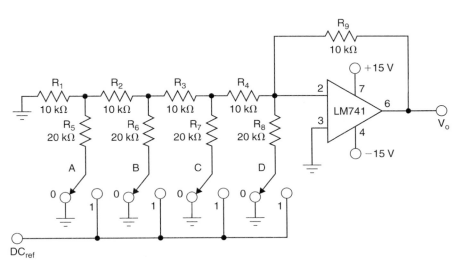

Figure 43-1

LAB PROCEDURE

PART 1 MANUAL BINARY INPUT

1. Construct the D/A circuit shown in Figure 43-1. Connect the +15 V and −15 V terminals of the power supply to the op-amp IC, and connect the +5-Vdc source to the DC_{ref} line of the circuit.

2. Set all four data switches so that 0 V is applied to all four inputs. Use the oscilloscope to measure the DC level at V_{out}, and record the value in the first line of Table 43-1. Repeat this step for all the logical combinations of binary inputs listed in the table.

3. From your results in Table 43-1, record the minimum and maximum values of output voltage.

   Note: Turn off the power and remove the oscilloscope from the circuit. You will be using the R-2R network and amplifier portion of the circuit in Part 2 of this project.

PART 2 BINARY COUNTER INPUT

1. Modify the R-2R D/A circuit of Figure 43-1 to replace the manual switches with a 4-bit binary counter (see the complete circuit in Figure 43-2).

2. Adjust the frequency of the function generator to 100 Hz. Adjust the oscilloscope to trigger internally on the CLK waveform. Adjust the oscilloscope to show three or four cycles of the waveform from V_{out}. Sketch the waveform at V_{out} in the space provided in Figure 43-3.

3. From the oscilloscope waveform of Step 2, determine the actual minimum and maximum output voltage. Record your findings on the Results Sheet.

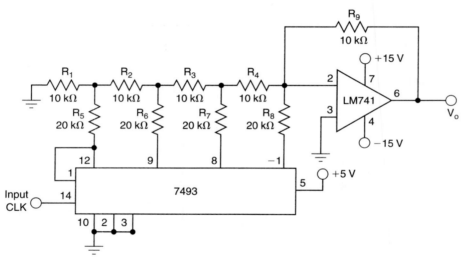

Figure 43-2

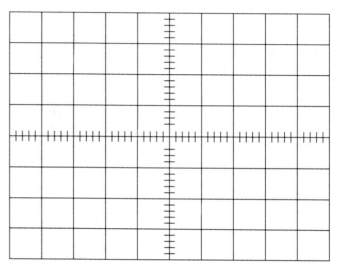

RESULTS SHEET

PROJECT 43

PART 1 MANUAL BINARY INPUT

Table 43-1

D_3	D_2	D_1	D_0	v_o
0	0	0	0	
0	0	0	1	
0	0	1	0	
0	0	1	1	
0	1	0	0	
0	1	0	1	
0	1	1	0	
0	1	1	1	
1	0	0	0	
1	0	0	1	
1	0	1	0	
1	0	1	1	
1	1	0	0	
1	1	0	1	
1	1	1	0	
1	1	1	1	

STEP 3

$V_{out(min)} =$ _____ $V_{out(max)} =$ _____

Questions

1. What is the resolution of the converter in this circuit?

2. Which binary input (D_0, D_1, D_2, or D_3) has the greatest amount of influence on the output voltage level?

PART 2 BINARY COUNTER INPUT

Figure 43-3

STEP 3

Measured $V_{out(min)} =$ _____

Measured $V_{out(max)} =$ _____

Questions

1. What is the actual resolution of the circuit?

2. How many input clock pulses are required to complete one cycle of the V_{out} waveform of Figure 43-3?

Critical Thinking for Project 43

1. Explain why it is important that the reference voltage for an R-2R D/A converter be as stable as possible.

2. A good way to test the operation of a D/A converter is by applying a binary counting sequence to the data inputs, as done in Part 2 of this project. Cite some circuit troubles and symptoms that you could detect with this kind of procedure.

PROJECT 44

A/D CONVERSION

A Prep Project

This project simulates a basic test setup for an 8-bit A/D converter. You will:

- Apply DC voltage levels to the input of the converter and note the corresponding binary output.
- Calculate the resolution of the circuit, and confirm the result by actual measurement.

Preparation

Read Frenzel, *Principles of Electronic Communication Systems*, Section 8-2.

Setup Procedure

1. Select **Prep Projects** from the **Projects** menu.

2. Select **Project 44   D/A Conversion.**

LAB PROCEDURE

For both parts of this project, the DC voltage source supplies the analog input voltage for the A/D converter circuit. In Part 1, the voltage range is 0 V to 6 V. In Part 2, the voltage range is −12 V to +12 V.
   The 8-bit digital display indicates the digital output of the A/D circuit.

PART 1   0-V TO 6-V INPUT

1. Adjust the DC voltage source for each of the values shown in Table 44-1. In each instance, record the corresponding binary output.

2. Experiment with the circuit to determine how much analog input voltage change is required for changing the binary output by a value of 1. Record your finding on the Results Sheet.

3. Adjust the DC voltage source to obtain the binary values cited in Table 44-2. Record the DC voltage source in each case.

PART 2   −12-V TO +12-V INPUT

The A/D converter in this part of the project is designed to handle positive and negative analog voltage levels. Such levels are often found in industrial monitoring and control systems.

1. Adjust the DC voltage source for each of the values shown in Table 44-3. In each instance, record the corresponding binary output.

2. Adjust the DC voltage source to obtain the binary values cited in Table 44-4. Record the DC voltage source in each case.

3. Experiment with the circuit to determine its maximum resolution. Record your finding on the Results Sheet.

A/D CONVERSION

A Hands-On Project

This project uses 8-bit A/D IC devices to help you become familiar with the A/D conversion process. You will:

- Construct the circuit.
- Apply data to the inputs and note the corresponding outputs.
- Calculate output values and confirm the results with actual measurement.

Preparation

1. Read Frenzel, *Principles of Electronic Communication Systems*, Section 8-2.

2. Complete the work for Prep Project 44.

Components and Supplies

1	Resistor, 10 kΩ
1	Potentiometer, 10 W
1	Capacitor, 100 pF
1	IC, 741 op amp
1	IC, ADC0804 8-bit A/D converter

The complete data sheet for this IC device can be found through the author's Web site at www.sweethaven.com/glencoe/commlab.

Equipment

1	Variable DC power supply
1	Function generator
1	Dual-trace oscilloscope
1	Digital voltmeter (optional)

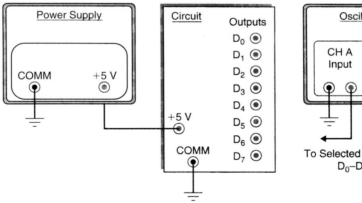
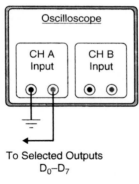

Figure 45-1

LAB PROCEDURE

The A/D device specified for this project (ADC0802) is intended for use with a microcomputer bus structure. However, it is used here as a stand-alone device. For this reason, latch up occasionally. The operation is easily and reliably restarted by momentarily shorting the connection between pins 3 and 5 to common. This is the purpose of the jumper in the circuit.

Special Note: The A/D device is sensitive to static discharge and voltage spikes that might occur when you construct the circuit or change any part of the circuit while DC power is applied to it. Do not apply power to the A/D circuit in this project until it is fully constructed. The DC voltage applied to V_{in} of the circuit (pin 6 on the A/D converter) should always be a positive voltage level between 0 and +5.1 Vdc.

1. Construct the A/D converter circuit as shown in Figure 45-2. Arrange the laboratory equipment as indicated in Figure 45-1.

2. Adjust the voltage applied to V_{in} to +5 V and note the DC voltage level at output D_7. If this output is not close to +5 V (logic 1 level), momentarily close the jumper connection between pin 3 and ground.

3. Adjust the voltage applied to V_{in} between 0 and +5 V, and note the level where output D_7 changes state. Record this level on the Results Sheet.

4. Set the input V_{in} to the series of voltage levels listed in Table 45-1. In each instance, use the oscilloscope to measure the digital output levels at D_0 through D_7. Record a 0-V level as 0, and a +5-V level as 1.

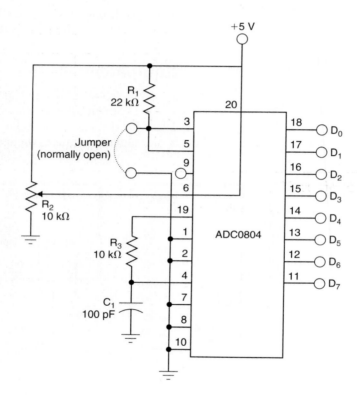

Figure 45-2

PROJECT
45

RESULTS SHEET

STEP 3

D_7 changes state when V_{in} is equal to _____ V.

Questions

1. Does the data in Table 45-1 indicate that the binary value of the output increases with the input voltage level? Explain your answer.

2. What is the theoretical resolution of the A/D converter in this project?

Table 45-1

v_i	D_7	D_6	D_5	D_4	D_3	D_2	D_1	D_0	Decimal Out
0.000									
0.500									
1.000									
1.500									
2.000									
2.500									
3.000									
3.500									
4.000									
4.500									
5.000									

Critical Thinking for Project 45

1. Describe a practical way to test the actual resolution of this A/D converter.

2. Define *quantizing error* and determine the maximum and average error for this circuit.

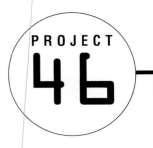

PULSE-CODE MODULATION

An Extended Project

This project simulates the operation of an 8-bit pulse-code modulator. You will:

- Apply various DC levels to the input of the modulator and note the sequence of output pulses.
- Determine the output pulse sequence for any amount of DC input.
- Closely observe the operation of a pulse-code modulator that has a low-frequency sinusoidal waveform applied at its input.

Preparation

Read Frenzel, *Principles of Electronic Communication Systems*, Section 8-4.

Setup Procedure

1. Select **Extended Projects** from the **Projects** menu.

2. Select **Project 46   Pulse-Code Modulation.**

LAB PROCEDURE

The pulse-code modulator in this project simulates the conversion of an analog input to a binary word. The lamps on the modulator indicate the input voltage level. The actual modulation is completed when the binary word is converted to a serial time-domain signal that appears on the oscilloscope display.

PART 1   VARIABLE DC INPUT

1. Adjust the DC voltage source between its two extremes, and note the response on the oscilloscope display. Record the number of pulses appearing on the screen when the input level is minimum and also when it is maximum.

2. Adjust the DC voltage source to the levels indicated in Table 46-1, and record the binary value of the waveform on the oscilloscope.

3. Set the DC voltage source to 18.2 V, and sketch the oscilloscope waveform on the grid in Figure 46-1.

PART 2   SIGNAL INPUT

The pulse-code modulator in this part of the project has a sinusoidal AC voltage applied to its input. While observing the operation of this circuit, answer the questions for Part 2 of the Results Sheet.

PULSE-WIDTH MODULATION

A Prep Project

This project simulates the action of a pulse-width modulator circuit. You will:

- Vary the DC input to a pulse-width modulator and measure the corresponding pulse width and voltage level at the output.
- Plot a performance curve showing output pulse width as a function of the input voltage.
- Determine the correspondence between the instantaneous voltage level of a sinusoidal input and the width of the pulses at the output.

Preparation

Read Frenzel, *Principles of Electronic Communication Systems*, Section 8-5.

Setup Procedure

1. Select **Prep Projects** from the **Projects** menu.

2. Select **Project 47   Pulse-Width Modulation.**

LAB PROCEDURE

PART 1   DC INPUT

Notice that the horizontal and vertical scaling for the oscilloscope appears in the upper right corner of the display. You will need these figures in order to determine the width and amplitude of the circuit's output pulse.

1. Vary the setting of the DC voltage source. Note how varying it affects the width of the pulse on the oscilloscope screen.

2. Set the DC voltage source to the values indicated in Table 47-1. Record the resulting output pulse width and amplitude.

3. Use the results of Table 47-1 to plot a graph showing how the pulse width changes with the amount of input voltage. Use the grid provided in Figure 47-1 on the Results Sheet.

PART 2   AUDIO INPUT

1. Adjust the output of the AC voltage source between its extreme settings. Note the response on the dual-trace oscilloscope display.

2. Set the AC voltage source for a 10 V_{p-p} signal. Sketch the two oscilloscope waveforms on the grid provided in Figure 47-2.

RESULTS SHEET

PART 1 DC INPUT

Table 47-1

Voltage In v_i	Pulse Width Out	Amplitude Out v_o
−8		
−6		
−4		
−2		
0		
2		
4		
6		
8		

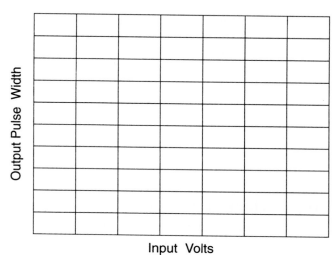

Figure 47-1

Questions

1. According to Figure 47-1, what is the amount of change in output pulse width per volt of change at the input?

2. According to Figure 47-1, what output pulse width would you expect upon applying −1 V to the input? +1 V?

3. What is the relationship between the amount of input voltage and the amplitude of the output signal?

211

PART 2 AUDIO INPUT

Questions

1. As the instantaneous input voltage rises, do the widths of the output pulses increase, decrease, or remain unchanged?

2. As the instantaneous input voltage falls, do the widths of the output pulses increase, decrease, or remain unchanged?

3. As the instantaneous input voltage rises, do the amplitudes of the output pulses increase, decrease, or remain unchanged?

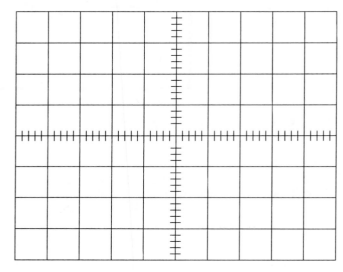

Figure 47-2

Critical Thinking for Project 47

1. Assuming that the sinusoidal input waveform of Figure 47-2 has a frequency of 100 kHz, determine the approximate sampling frequency of the modulator.

2. Explain how it is possible to use an AM diode detector circuit as a PWM demodulator.

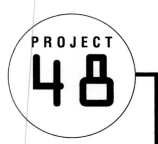

PROJECT 48

PULSE-WIDTH MODULATION

A Hands-On Project

This project uses timer circuits to obtain pulse-width modulation. You will:

- Construct the circuit.
- Confirm the operation of the two basic parts of the circuit.
- Apply an AC waveform and observe the resulting pulse-width output.

Preparation

1. Read Frenzel, *Principles of Electronic Communication Systems*, Section 8-5.

2. Complete the work for Prep Project 47.

Parts and Materials

2	Resistors, 2.7 kΩ
1	Resistor, 10 kΩ
1	Capacitor, 1 pF
1	Capacitor, 10 nF
1	Capacitor, 33 nF
1	Capacitor, 100 nF
2	ICs, LM555 timer

Equipment

1	DC power supply
1	Function generator
1	Dual-trace oscilloscope

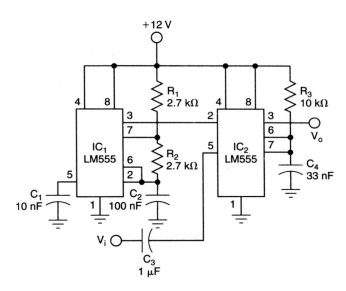

Figure 48-1

LAB PROCEDURE

1. Construct the circuit in Figure 48-1. Connect the oscilloscope to monitor pin 3 of IC1 on one trace and V_{in} (pin 3 of IC2) on the other trace. Do not connect the function generator to input V_{in} at this time.

2. Adjust the sweep rate to 0.1 ms and set the vertical sensitivity of both channels to 5 V/div. Sketch the display in Figure 48-2. Also determine the frequency of both waveforms, and the duration of the positive portion of both waveforms. Record your data on the Results Sheet.

Note: The waveform from IC1 is the output of a free-running oscillator. The waveform from IC2 is the output of a monostable multivibrator that is triggered by the negative-going edge of the oscillator's output pulse.

3. Set the oscilloscope to monitor V_{in} on one channel and V_{out} on the other. Trigger the sweep from the signal at V_{out}. Connect the function generator to input V_{in} of the circuit, and adjust the generator for a sinusoidal output of about 100 Hz, at 4 V_{p-p}. Leave the sweep rate at 0.1 ms/div.

4. Use the fine-frequency adjustment on the function generator to trim the frequency so that one full cycle of the waveform appears stationary on the oscilloscope display. Note how the duration of the pulses at V_{out} varies with the instantaneous voltage level at V_{in}. Make a rough sketch of the display in Figure 48-3. Record the minimum and maximum pulse widths you observe at V_{out}.

Name _____ Date _____

RESULTS SHEET

STEP 2

Oscillator frequency = _____

V_{out} frequency = _____

Oscillator positive duration = _____

V_{out} positive duration = _____

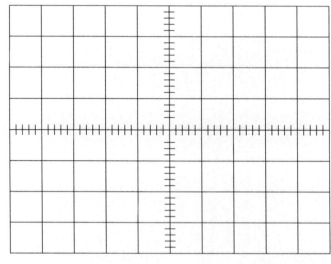

Figure 48-2

STEP 4

Minimum pulse width = _____

Maximum pulse width = _____

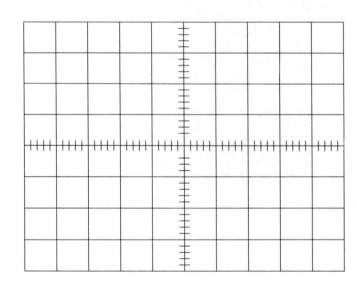

Figure 48-3

Questions

1. According to the data from Step 4 and the display in Figure 48-3, does the output pulse width increase or decrease as the input AC waveform goes more positive?

2. Which IC in this circuit represents the clock section of the PWM? The modulator section?

Critical Thinking for Project 48

1. The operation of this particular pulse-width modulator depends on a voltage-controlled monostable multivibrator. Explain how this dependence relates to the circuit in this project.

2. Describe the type of circuit that is commonly used for demodulating a PWM waveform.

THE COMPANDER BLOCK

A commsim Project

This project demonstrates the operation of a compander/expander system. You will:

- Note the basic operation of the compander and expander blocks.
- Vary the type of signal input waveform and note the changes in the companded and expanded versions.

Setup Procedure

1. Start commsim on your computer.

2. Make sure that your communication lab CD-ROM is in the computer's CD drive.

3. Open the **commsim** directory on the CD-ROM.

4. Select **Project_49.vsm.**

5. Look for a worksheet diagram that is similar to the one shown in Figure 49-1.

- The Signal Source is a waveform generator that you can set for triangle, sawtooth, and rectangular waveforms. This is the signal source that is companded and then later expanded and restored.
- The Compander block is responsible to doing the data compression operation.
- The plotter for Compander Waveforms displays the signal source and the output of the compander.
- The Expander block restores the original signal. In the real world, of course, this operation takes place at different location in space or time.
- The plotter labeled Expander output shows the output of the Expander block.
- Ideally, the signal input on the Compander Waveforms plot should be identical to the waveform on the Expander Output plot.

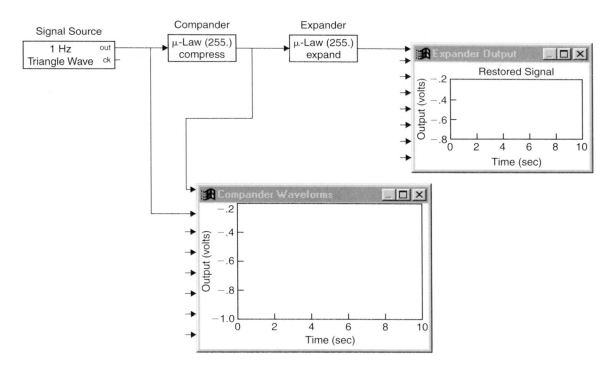

Figure 49-1

LAB PROCEDURE

1. Start the simulation.

2. Observe the behavior with the default settings.

3. Sketch the compander waveforms (input signal and companded signal) in Figure 49-2.

4. Double-click the Signal Source block to access the Waveform Generator Properties dialog box.

5. Set the Waveform Type to Square Wave.

6. Close the dialog box and run the simulation.

7. Describe the differences, if any, you can see between the original rectangular waveform and its companded version.

8. Use the dialog box for the Signal Source to set up the Sawtooth Wave.

9. Run the simulation and note the results. Sketch the waveforms from the Compander Waveforms plotter on Figure 49-3.

PROJECT
49

RESULTS SHEET

STEP 3

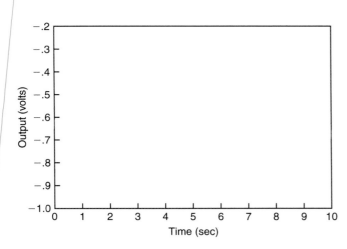

Figure 49-2

STEP 9

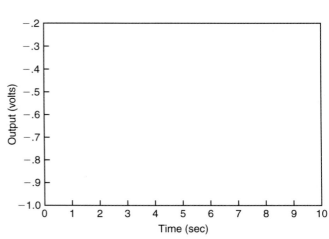

Figure 49-3

STEP 7

Note any substantial differences between the input rectangular waveform and its companded version:

_____

_____

Critical Thinking for Project 49

Explain why a companded rectangular waveform does not look substantially different from the expanded version.

SAMPLE-AND-HOLD CIRCUIT

A multiSIM Project

Sample-and-hold (S/H) circuits grab and hold an analog voltage value. In this project you will:

- Observe the operation of a simplified sample-and-hold (S/H) circuit.
- Determine sampling rates and hold intervals.
- Observe the effects of changing the analog input signal.

Preparation

Read Frenzel, *Principles of Electronic Communication Systems*, Section 8-2.

Setup Procedure

1. Start multiSIM on your computer.

2. Make sure that your communication lab CD-ROM is in the computer's CD drive.

3. Open the **multiSIM** directory on the CD-ROM.

4. Select **Project_50.msm.**

5. Look for a worksheet diagram that is similar to the one shown in Figure 50-1.

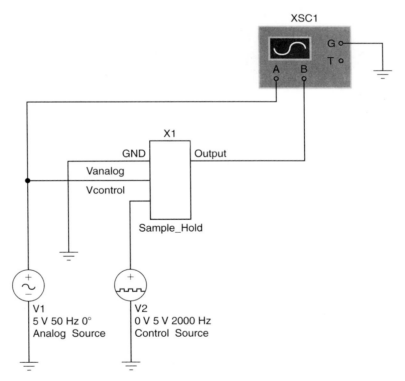

Figure 50-1

LAB PROCEDURE

PART 1 USING THE DEFAULT SETTINGS

1. Expand the oscilloscope instrument and start the simulation with the project's default settings.

2. Sketch the essential features of the waveforms on the grid provided in Figure 50-2.

3. Record the sampling rate and the analog input voltage level.

4. Superimpose the two waveforms by setting the Y position for channels A and B to 0 V. Notice how the output waveform samples the analog input and "holds" it for a fixed period of time. Record the hold interval.

PART 2 CHANGING THE ANALOG FREQUENCY

1. Turn off the simulation and double-click the analog source (V_1) to show the AC voltage dialog box.

2. Change the frequency setting from the default value of 100 Hz to a higher value of 200 Hz.

3. Close the dialog box, make sure that the oscilloscope instrument is fully expanded, and start the simulation.

4. Sketch the essential features of the waveforms on the grid provided in Figure 50-3.

5. Record the sampling rate and hold interval.

6. Repeat Steps 1 to 3, setting the analog source frequency to 50 Hz.

7. Sketch the essential features of the waveforms on the grid provided in Figure 50-4.

8. Record the sampling rate and hold interval.

PROJECT
50

RESULTS SHEET

PART 1 USING THE DEFAULT SETTINGS

STEP 2

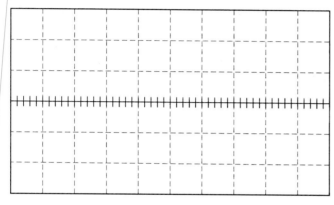

Figure 50-2

PART 2 CHANGING THE ANALOG FREQUENCY

STEP 3

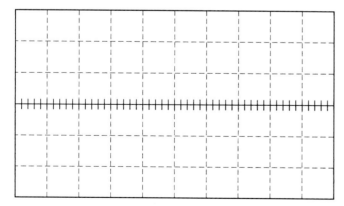

Figure 50-3

STEP 3

Sampling rate = _____

Analog voltage input level = _____

STEP 4

Hold interval = _____

Questions

1. Why does the output waveform have a stair-step appearance?

2. What is the mathematical relationship between the sampling rate and the hold interval?

STEP 4

Sampling rate = _____

Hold interval = _____

STEP 7

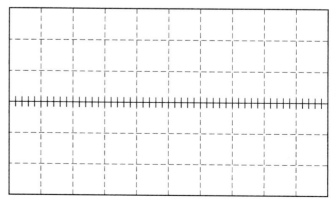

Figure 50-4

STEP 8

Sampling rate = _____

Hold interval = _____

Questions

1. Are your values for the sampling rate and hold intervals roughly the same for each part of this project? If not, you should go over your procedures very carefully.

2. Which output waveform—the one in Figure 50-1, 50-2 or 50-3—most accurately represents its analog input voltage waveform?

Critical Thinking for Project 50

1. In theory, it is possible to increase the accuracy of the output waveform by increasing the sampling rate. Explain why this so.

2. You were not instructed to change the sampling rate of the circuit. This is because the value of the hold capacitor inside the S/H circuit is selected for the 2-kHz sampling rate. You demonstrated the same results, however, by changing the analog input frequency. Explain how the *ratio* of the analog input frequency to the sampling rate determines the quality of the output waveform.

3. Explain why the sampling rate is practically of no consequence when sampling and holding DC levels.

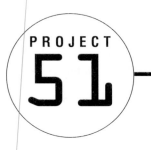

FREQUENCY CONVERTER

A Prep Project

The circuit in this project is a frequency converter, or mixer. The input to the circuit is two different frequencies, f_s and f_o. The output is four different frequencies: f_s, f_o, $f_s + f_o$, and $f_s - f_o$. This project gives you a chance to select the amplitude and frequency of waveforms f_s and f_o that are applied to the converter circuit. You can then confirm that the output of the circuit consists of the four just cited: the sum, difference, and two original frequencies.

In this project you will:

- Set the frequency and voltage levels for the two inputs of a typical frequency converter circuit.
- Use a tuned circuit and AC voltmeter to determine the frequencies and amplitudes of the signals at the output of a converter circuit.
- Use a simulated spectrum analyzer to confirm the presence of four different frequencies at the output of a converter circuit.

Preparation

Read Frenzel, *Principles of Electronic Communication Systems*, Section 8-3.

Setup Procedure

1. Select **Prep Projects** from the **Projects** menu.

2. Select **Project 51   Frequency Converter.**

LAB PROCEDURE

The procedures you will be using for setting the input amplitudes and frequencies are the same for both Parts 1 and 2 of the project. The difference between the two parts is the type of simulated instruments you will be using for measuring the output frequencies and amplitudes.

In Part 1, the output of the converter circuit is fed to a tuned circuit. The tuned circuit is connected to a digital AC voltmeter. So, as you tune the circuit to the four separate output frequencies, you will see readings on the voltmeter rising and falling across peak voltage levels.

In Part 2, you will replace the tuned circuit and AC voltmeter with a spectrum analyzer. In this case, an oscilloscope display clearly shows the four output frequencies and their individual voltage levels.

PART 1   TUNED OUTPUT FILTER

1. Set the amplitudes for both RF generators to 12.6 V. Set the output of the RF generator (labeled f_s) to 50.0 MHz. Set the output of the second RF generator (labeled f_o) *to* 5.0 MHz. Record on the Results Sheet the four frequencies you can expect to find at the output of the mixer circuit.

2. Adjust the tuned filter to select the four frequencies from the mixer circuit. Locate the peaks by noting the maximum voltage levels on the digital AC voltmeter. List the four peak frequencies and their voltage levels in Table 51-1 of the Results Sheet.

PART 2 SPECTRAL ANALYSIS

Note: The horizontal grid on the spectrum analyzer display is scaled at 10 MHz/div. The vertical grid is scaled at 10 V/div.

1. Set the amplitudes and frequency for both RF generators specified for Part 1 of this project:

   RF generator—f_s
       Amplitude: 12.6 V   Frequency: 50.0 MHz
   RF generator—f_o
       Amplitude: 12.6 V   Frequency: 5.0 MHz

2. Sketch the spectrum you see on the analyzer display. Use the grid in Figure 51-1 on the Results Sheet. Label the peaks as f_s, f_o, $f_s - f_o$, and $f_s + f_o$.

3. Determine the frequency and amplitude of each of the peaks on the display. Record your findings in Table 51-2.

4. Set up your own values of frequency and voltage for the inputs to the circuit. Record your values on the Results Sheet.

5. Sketch the spectrum you see on the analyzer display. Use the grid in Figure 51-2 on the Results Sheet.

6. Determine the frequency and amplitude of each of the peaks on the display. Record your findings in Table 51-3.

PROJECT

51

RESULTS SHEET

PART 1 TUNED OUTPUT FILTER

STEP 1

$f_s = $ _____ $f_o = $ _____

$f_s + f_o = $ _____

$f_s - f_o = $ _____

Table 51-1

Peak	Frequency	Amplitude
#1		
#2		
#3		
#4		

Questions

1. Do the results in Step 2 confirm the results anticipated in Step 1?

2. If frequency f_s is increased, what should you do to the frequency f_o in order to maintain the same value for $f_s - f_o$?

PART 2 SPECTRAL ANALYSIS

Table 51-2

Peak	Frequency	Amplitude
#1		
#2		
#3		
#4		

Figure 51-1

STEP 2

RF generator—f_s:

  Amplitude = _____

  Frequency = _____

RF generator—f_o:

  Amplitude = _____

  Frequency = _____

Questions

1. Which one of the four peaks is most sensitive to the amplitude of input f_s?

2. Which one of the four peaks is most sensitive to the amplitude of input f_o?

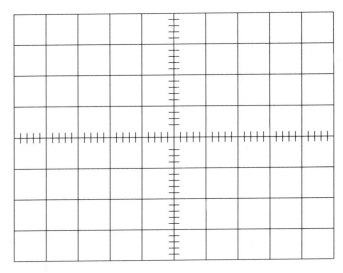

Figure 51-2

Table 51-3

Peak	Frequency	Amplitude
#1		
#2		
#3		
#4		

Critical Thinking for Project 51

1. Explain how using a tuned circuit to determine the frequencies at the output of a frequency converter is related to the use of a tuned circuit to select one radio station from all others.

2. Suppose frequency sources f_s and f_o were "ganged" so that the $f_s - f_o$ output always remained at the same frequency. Describe how this would affect the appearance of the four peaks on the spectrum analyzer as you varied one of the frequency sources.

3. Explain why the tuned filter of Part 1 is to be removed before attaching the output of the frequency converter to the spectrum analyzer in Part 2.

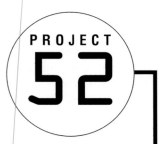

FREQUENCY CONVERTER

PROJECT 52

A Hands-On Project

This project uses a germanium diode as the nonlinear element of a frequency converter, or mixer. You will:

- Construct the circuit.
- Apply frequencies to be mixed.
- Verify the presence of the sum and difference frequencies at the output of the circuit.

Preparation

1. Read Frenzel, *Principles of Electronic Communication Systems*, Section 8-3.

2. Complete the work for Prep Project 38.

Components and Supplies

2	Resistors, 1 kΩ
2	Resistors, 1.5 kΩ
1	Diode, 1N34 or equivalent
1	455-kHz ceramic filter

Equipment

1	Dual-trace oscilloscope
2	Function generators
1	Frequency counter (optional)

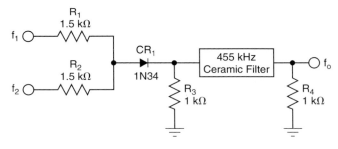

Figure 52-1

LAB PROCEDURE

1. Construct the mixer circuit shown in Figure 52-1. Connect one frequency generator to input f_1 of the circuit, and connect the second to input f_2. Set the output amplitude of both frequency generators to about 6 V_{p-p}. Use sinusoidal waveform outputs for both generators.

2. Using the values for f_1 given in Table 52-1 on the Results Sheet, calculate the values of f_2 such that $f_1 + f_2 = 455$ kHz. Enter your results in the f_2 column.

3. Apply each combination of f_1 and f_2 in Table 52-1 to the inputs of the circuit. Use the oscilloscope (or frequency counter) to verify the presence of a 455-kHz signal at the output of the circuit.

4. Using the values for f_1 given in Table 52-2, calculate the values of f_2 such that $f_1 - f_2 = 455$ kHz. Enter your results in the f_2 column.

5. Apply each combination of f_1 and f_2 in Table 52-2 to the inputs of the circuit. Use the oscilloscope (or frequency counter) to verify the presence of a 455-kHz signal at the output of the circuit.

6. Set input f_1 to 455 kHz, and vary the frequency of f_2 to 100 kHz above and 100 kHz below f_1. Verify that 455 kHz remains as an output frequency.

7. Set input f_2 to 455 kHz, and vary the frequency of f_1 to 100 kHz above and 100 kHz below f_2. Once again, verify that 455 kHz remains as an output frequency.

RESULTS SHEET

Table 52-1

Input f_1 (kHz)	Input f_2 (kHz)	Output f_o (kHz)
100		455
150		455
200		455
250		455
300		455

Table 52-2

Input f_1 (kHz)	Input f_2 (kHz)	Output f_o (kHz)
700		455
750		455
800		455
850		455
900		455

STEP 6

Is 455 kHz present at the output for all values of f_2?

_____

STEP 7

Is 455 kHz present at the output for all values of f_1?

_____

Questions

1. What is the function of the portion of the circuit composed of R_1, R_2, and C_{R1}?

2. What will be the output frequency from this circuit if f_1 and f_2 are both set to 455 kHz?

Critical Thinking for Project 52

1. Cite the similarities in the operation of an AM modulator and a frequency converter.

2. The frequency 455 kHz is commonly found in commercial AM radio receivers. What role does it play in these receivers?

Experimental Notes and Calculations

SIMPLE AM RECEIVER

A Hands-On Project

This project uses an LM386 audio amplifier as a primitive AM receiver that is similar in principle to the crystal radios used in the early days of radio technology. You will:

- Construct the circuit.
- Make appropriate antenna and ground connections.
- Identify the carrier frequency of the broadcast being heard.

Preparation

Read Frenzel, *Principles of Electronic Communication Systems*, Section 8-1.

Components and Supplies

1   Capacitor, 100 nF
1   Capacitor, 10 μF
1   IC, LM386 audio power amplifier
1   8 Ω permanent-magnet speaker or earphone
      10 ft of copper wire, about 18 gauge

Equipment

1   DC power supply

The circuit behaves as a detector and audio amplifier, with the detection taking place at PN junctions located at the input terminal of the IC. Since there are no provisions for tuning the circuit, it will respond to the strongest AM broadcast signal in your area.

LAB PROCEDURE

1. Construct the circuit shown in Figure 53-1.

2. Set up the antenna. Use the long wire as the antenna, stretching it out as long as possible. Make sure that you have a good electrical connection at the amplifier end.

3. Make the ground connection. If an actual water pipe connection is not available, try making a connection to electrical conduit or the metallic faceplate of an electrical outlet. Be sure to connect the COMM of the power supply to this ground as well.

4. Describe your observations of this project on the Results Sheet.

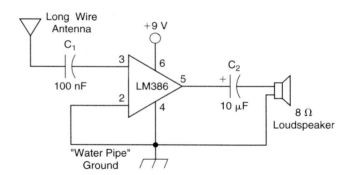

Figure 53-1

PROJECT

53

RESULTS SHEET

STEP 4

_____

_____

_____

_____

_____

Questions

1. What is the carrier frequency for the broadcast you are receiving with this circuit? (*Hint:* Local newspapers usually publish the broadcast frequency of the AM and FM stations in your area.)

2. What would be the advantage of replacing C_1 with a bandpass filter that could be adjusted for resonant frequencies between 550 kHz and 1650 kHz?

Critical Thinking for Project 53

1. Research and describe the operation of a simple crystal radio receiver.

2. Explain why a long antenna and good electrical connection to ground are essential for the operation of this circuit.

PROJECT 54

SUPERHETERODYNE RECEIVERS

An Extended Project

This project uses simulated electronic instruments and an interactive block diagram to demonstrate the flow of signals through a typical superheterodyne AM radio receiver. In this project you will:

- Trace the flow of signals through the receiver.
- Identify the RF and audio frequencies at critical test points.
- Identify the function of certain parts of the receiver system.

Preparation

Read Frenzel, *Principles of Electronic Communication Systems*, Section 8-5.

Setup Procedure

1. Select **Extended Projects** from the **Projects** menu.

2. Select **Project 54   Superheterodyne Receivers**.

LAB PROCEDURE

Before beginning the actual lab procedure, notice that the project uses two different test instruments: an AM RF generator and a set of frequency counters that measure RF and audio signals.

You can adjust the output of the RF generator by means of the RF frequency slide control. The audio modulating frequency is fixed at 540 Hz, and the percent of modulation is fixed at 100 percent. Take a moment to experiment with this control.

The measuring instruments located in the upper-right corner of the screen measure radio and audio frequencies at the same time. These two are always connected to the same test point. The RF counter indicates the RF frequency at the test point, and the audio counter indicates the audio frequency at the same point. The only electrical difference between the two counters is that the audio counter has a low-pass filter connected in series with its input.

1. Locate the tuning and band controls on the interactive block diagram. Adjust the tuning control for 1454 kHz.

   You have just tuned your simulated AM radio receiver for a station located at 1454 kHz.

2. Likewise, adjust the AM RF generator for 1454 kHz.

3. Move the mouse pointer to test point TP 1 on the interactive block diagram. It is found at the antenna.

   Notice that the pointer changes from an arrow to a rectangle. This change indicates that you have touched TP 1 with the probe for the read-out instruments. Record on the Results Sheet the frequencies you find at TP 1.

4. Move the mouse pointer to TP 2, and record the radio and audio frequencies you find at that point. Then move the pointer to TP 3 and TP 4. Record your figures on the Results Sheet. Be prepared to describe why the readings at these three points differ.

5. Observe and record the frequencies at TP 5 and TP 6. Use the spaces provided on the Results Sheet. Be prepared to describe why the readings at these two points differ.

6. Observe and record the frequencies at TP 7.

7. Make sure that the tuning control on the interactive block diagram is still set for 1454 kHz. Adjust the AM RF generator for 1200 kHz. Observe the frequencies at all the test points, and be prepared to explain why they are so different from the results obtained for the first part of this project.

Name _____ Date _____

RESULTS SHEET

STEP 3

TP 1 radio frequency = _____

TP 1 audio frequency = _____

STEP 4

TP 2 radio frequency = _____

TP 2 audio frequency = _____

TP 3 radio frequency = _____

TP 3 audio frequency = _____

TP 4 radio frequency = _____

TP 4 audio frequency = _____

According to the block diagram, TP 3 indicates the output of the

_____

_____

According to the block diagram, TP 4 indicates the output of the

_____

_____

STEP 5

TP 5 radio frequency = _____

TP 5 audio frequency = _____

TP 6 radio frequency = _____

TP 6 audio frequency = _____

According to the block diagram, TP 6 indicates the output of the

_____

_____

STEP 6

TP 7 radio frequency = _____

TP 7 audio frequency = _____

According to the block diagram, TP 7 indicates the output of the

_____

_____

Questions

1. How do you account for the difference in the radio frequency between TP 2 and TP 4, as recorded in Step 4?

2. Why is there such a difference between the readings at TP 5 and TP 6, as recorded in Step 5 of this project?

3. Why is there no RF signal at TP 7, as recorded in Step 6?

4. Which instrument in this circuit is playing the role of a commercial AM broadcast transmitter?

Critical Thinking for Project 54

1. Suppose that the tuning of the receiver (interactive block diagram) is set for 1230 kHz. Further suppose that the AM RF generator is set for 1430 kHz on the AM band. What will be the signals at TP 1? Describe what you should see at TP 2. Explain these results.

2. Explain how the frequency at TP 3 is always made to be 455 kHz higher than the frequency at TP 1.

3. Explain why there is no audio signal at TP 7 whenever the tuning of the receiver and the frequency of the AM RF generator differ by 5 kHz or more.

AM/FM RADIO RECEIVER

An Extended Project

This project uses simulated electronic instruments and an interactive block diagram to demonstrate the flow of signals through a typical AM/FM radio receiver. In this project you will:

- Trace the flow of signals through the receiver when it is operating as an AM receiver.
- Trace the flow of signals through the receiver when it is operating as an FM receiver.
- Identify the function of certain parts of the receiver system.

Preparation

Read Frenzel, *Principles of Electronic Communication Systems*, Section 8-5.

Setup Procedure

1. Select **Extended Projects** from the **Projects** menu.

2. Select **Project 55   AM/FM Radio Receiver.**

LAB PROCEDURE

Before starting the actual lab procedure, notice that the project uses two different test instruments: an AM/FM RF generator and a frequency counter that measures radio and audio signals at the same time.

You can adjust the output of the RF generator by means of the radio frequency slide control. And you can determine whether the range of frequencies are in the commercial AM or FM broadcast band by pressing the toggle switch on the generator. The audio modulating frequency is fixed at 400 Hz, and the percent of modulation is fixed at 100 percent. Take a moment to experiment with the slide adjustment and toggle the pushbutton on the RF generator.

The measuring instruments located in the upper-right corner of the screen measure radio and audio frequencies at the same time. These two are always connected to the same test point. The RF counter indicates the radio frequency at the test point, and the audio counter indicates the audio frequency at the same

point. The only electrical difference between the two counters is that the audio counter has a low-pass filter connected in series with its input.

Two kinds of block diagrams are visible on the screen at all times. The interactive block diagram clearly shows sections of the AM/FM receiver circuit and the test points for this project. This block diagram is too large to fit on the screen, but you can view all portions by scrolling the diagram to the left or right. Use the horizontal scroll bar near the bottom of the screen for this purpose.

A smaller version of the same block diagram is located near the top of the display. This is the navigation block diagram. Its purpose is to indicate the section of the larger diagram you are viewing at the moment. The blue rectangle on the navigation block diagram indicates where you are on the larger interactive diagram.

Finally, you have access to a small block diagram that shows how the instruments and circuit are interconnected. You can turn this block diagram on and off by clicking the block diagram button located near the upper-right corner of the display.

During this first section of the lab procedure, you will work with the receiver and test instruments in an AM mode.

1. Locate the tuning and band controls on the interactive block diagram. Click the band control for the AM band, and adjust the tuning control for 810 kHz. You have just tuned your simulated AM/FM radio receiver for a station located at 810 kHz on the AM band.

2. Likewise, adjust the AM/FM RF generator for 810 kHz of the AM band.

3. Locate test point TP 1 on the interactive block diagram. It is found at the antenna. Scroll the diagram to the left or right if necessary for seeing TP 1. Move the mouse pointer to TP 1. Notice that the pointer changes from an arrow to a rectangle. This change indicates that you have touched TP 1I with the probe for the read-out instruments. Note the frequencies. If these

output frequencies are not 810 kHz and 400 Hz, check your setup to this point in the project.

4. Move the mouse pointer to TP 2, and record the radio and audio frequencies you find at that point. Use the spaces provided on the Results Sheet. Then move the pointer to TP 3 and record the results. Be prepared to describe why the readings at these two points differ.

5. Observe and record the frequencies at TP 4 and TP 5. Use the spaces provided on the Results Sheet.

6. Observe and record the frequencies at TP 6, TP 7, and TP 8.

7. Observe and record the frequencies at TP 9 and TP 10. Be prepared to describe why the readings at these two points differ.

8. Observe and record the frequencies at TP 11 and TP 12. Be prepared to describe why the readings at these two points are different.

9. Observe and record the frequencies at TP 14 and TP 16.

10. Observe and record the output at TP 17.

During this second phase of the lab procedure, you will work with the receiver and test instruments in the FM mode.

11. Set the tuning and band controls on the interactive block diagram for 90 MHz on the FM band. Then tune the output of the AM/FM RF generator to the same frequency.

12. Observe and record the frequencies at TP 1.

13. Observe and record the frequencies at TP 2 and TP 3. Use the spaces provided on the Results Sheet, and be prepared to explain any differences.

14. Observe and record the frequencies at TP 4 and TP 5.

15. Observe and record the frequencies at TP 6, TP 7, and TP 8.

16. Observe and record the frequencies at TP 9 and TP 10. Be prepared to describe why the readings at these two points differ from those found for the AM mode in Step 7.

17. Observe and record the frequencies at TP 13.

18. Observe and record the output at TP 15.

RESULTS SHEET

AM Mode

STEP 4

TP 2 radio frequency = _____

TP 2 audio frequency = _____

TP 3 radio frequency = _____

TP 3 audio frequency = _____

According to the block diagram, TP 2 indicates the output of the

_____

_____

According to the block diagram, TP 3 indicates the output of the

_____

_____

STEP 5

TP 4 radio frequency = _____

TP 4 audio frequency = _____

TP 5 radio frequency = _____

TP 5 audio frequency = _____

STEP 6

TP 6 radio frequency = _____

TP 6 audio frequency = _____

TP 7 radio frequency = _____

TP 7 audio frequency = _____

TP 8 radio frequency = _____

TP 8 audio frequency = _____

STEP 7

TP 9 radio frequency = _____

TP 9 audio frequency = _____

TP 10 radio frequency = _____

TP 10 audio frequency = _____

STEP 8

TP 11 radio frequency = _____

TP 11 audio frequency = _____

TP 12 radio frequency = _____

TP 12 audio frequency = _____

According to the block diagram, TP 11 indicates an input to a(n)

_____

_____

STEP 9

TP 14 radio frequency = _____

TP 14 audio frequency = _____

TP 16 radio frequency = _____

TP 16 audio frequency = _____

STEP 10

TP 17 radio frequency = _____

TP 17 audio frequency = _____

Results Sheet for Project 55 (*continued*)

According to the block diagram, TP 17 indicates the
output of the

_____

_____

FM Mode

STEP 12

TP 1 radio frequency = _____

TP 1 audio frequency = _____

STEP 13

TP 2 radio frequency = _____

TP 2 audio frequency = _____

TP 3 radio frequency = _____

TP 3 audio frequency = _____

STEP 14

TP 4 radio frequency = _____

TP 4 audio frequency = _____

TP 5 radio frequency = _____

TP 5 audio frequency = _____

STEP 15

TP 6 radio frequency = _____

TP 6 audio frequency = _____

TP 7 radio frequency = _____

TP 7 audio frequency = _____

TP 8 radio frequency = _____

TP 8 audio frequency = _____

STEP 16

TP 9 radio frequency = _____

TP 9 audio frequency = _____

TP 10 radio frequency = _____

TP 10 audio frequency = _____

STEP 17

TP 13 radio frequency = _____

TP 13 audio frequency = _____

STEP 18

TP 15 radio frequency = _____

TP 15 audio frequency = _____

Questions

1. Why is there such a great difference between the readings at TP 2 and TP 3, as recorded in Step 4 of this project?

2. How do you account for the difference in the radio frequency between TP 4 and TP 5, as recorded in Step 5?

3. Why is there such a difference between the readings at TP 9 and TP 10, as recorded in Step 7 of this project?

4. Why is there no audio signal at TP 11, as recorded in Step 8?

5. Why is there no RF signal at TP 17, as recorded in Step 10?

6. Why is there such a great difference between the readings at TP 2 and TP 3, as recorded in Step 13?

7. Why are the readings for the same two points, TP 9 and TP 10, different for the AM mode (Step 7) and FM mode (Step 16)?

Critical Thinking for Project 55

1. Suppose the tuning on the receiver (interactive block diagram) is set for 1230 kHz on the AM band. Further suppose the AM/FM RF generator is set for 1430 kHz on the AM band. What will be the signals at TP 1? Describe what you should see at TP 2 and TP 3 and explain these results.

2. Suppose the tuning on the receiver and the RF generator are both set for the FM band, but at vastly different frequencies. Describe both the RF and audio signals you should find at TP 1, TP 2, TP 3, and TP 4.

PROJECT 56

SIGNALS AND NOISE

A commSIM Project

Preparation

Read Frenzel, *Principles of Electronic Communication Systems*, Section 9-5.

Setup Procedure

1. Start multiSIM on your computer.

2. Make sure that your communication lab CD-ROM is in the computer's CD drive.

3. Open the **commSIM** directory on the CD-ROM.

4. Select **Project_56.vsm.**

5. Look for a worksheet diagram that is similar to the one shown in Figure 56-1.

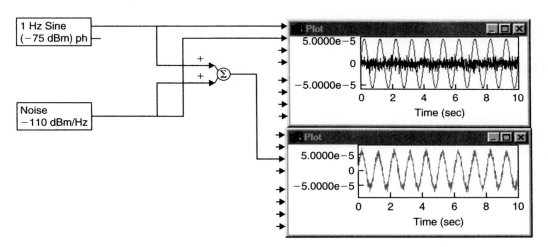

Figure 56-1

LAB PROCEDURE

1. Run the simulation with the following default values and note the waveforms on the plot instruments:

   - The test signal is a clean, 1-Hz sinusoidal waveform at −75 dBm. See the red trace.
   - The noise signal is set at −110 dBm/Hz. See the blue trace.
   - The circuit sums the test and noise signals to produce the resulting "noisy" sine wave. See the green trace.

2. Estimate the peak-to-peak voltage levels for the signal (V_s) and the noise (V_n) as they appear on the upper plotter. Use these values to calculate the signal-to-noise ratio (S/N). Sketch the noisy waveform (green trace) in Figure 56-2.

3. Click the Noise device to access the Noise Properties dialog box.

4. Increase the noise level to −100 dBm.

5. Close the dialog box and run the simulation.

6. Estimate and record the signal and noise voltage levels. Calculate the S/N ratio and sketch the noisy waveform in Figure 56-3.

7. Change the noise level to −120 dBm.

8. Estimate and record the signal and noise voltage levels. Calculate the S/N ratio.

PROJECT
56

RESULTS SHEET

STEP 2

$V_s =$ _____ $V_n =$ _____

$S/N =$ _____

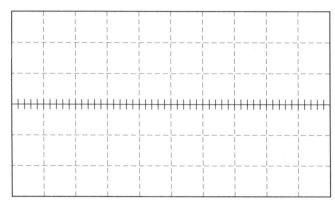

Figure 56-2

STEP 6

$V_s =$ _____ $V_n =$ _____

$S/N =$ _____

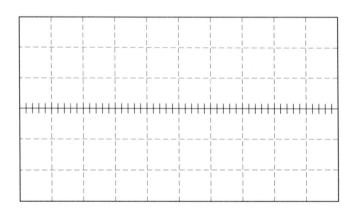

Figure 56-3

STEP 8

$V_s =$ _____ $V_n =$ _____

$S/N =$ _____

Questions

1. What effect does increasing the amount of noise have upon the S/N ratio?

2. Which is preferred in a communication system—a high S/N ratio or a low S/N ratio?

Critical Thinking for Project 56

1. The noise component in this project can generate noise at levels shown in dBm per hertz. Explain the significance of a frequency component (Hz) in the measurement.

2. Explain how simple filter circuits can improve the S/N ratio of a noisy signal.

PROJECT 57

DATA MULTIPLEXER

A Prep Project

This project uses an interactive block diagram as a tool for studying the flow of data through an eight-channel data multiplexer/demultiplexer system. You will:

- Observe simulated analog data sources at the inputs of an eight-channel multiplexer.
- Relate the binary select code to the input channel that is being transmitted.
- Observe the flow of data through the single data line.
- Relate the binary select code to the output channel that is receiving data.

Preparation

Read Frenzel, *Principles of Electronic Communication Systems*, Section 10-3.

Setup Procedure

1. Select **Prep Projects** from the **Projects** menu.

2. Select **Project 57   Data Multiplexing.**

LAB PROCEDURE

1. This test setup consists of a 3-bit binary code generator, an interactive block diagram of an eight-channel multiplexer/demultiplexer, and a status board that indicates the nature of the signal at selected test points.

   Set the binary generator for its manual mode of operation. Set the data switches to the eight different combinations shown in Table 57-1. In each case, click the Load button to transfer the data to the select lines of the multiplexer and demultiplexer.

2. For each of the channel-select codes in Table 57-1, indicate which data input channel is actually being transmitted (Ch 1, Ch 2, Ch 3, and so on). Verify your conclusion by moving the mouse pointer to TP 1 and recording the channel number found at that point.

3. Also, for each channel-select code in Table 57-1, indicate which data output channel is actually receiving data.

4. Once again, set the data switches on the binary code generator to the eight combinations of codes shown in Table 57-2. In this case, however, move the mouse pointer to each of the data output terminals. Mark channels receiving no data with an X. Mark the channel that is receiving data with the channel number being received (Ch 1, Ch 2, Ch 3, and so on).

5. Click the mode button on the binary code generator for the count-up mode of operation. Click the rate button for the slow counting speed. Use the mouse pointer to observe each data input terminal, and answer Question 1 on the Results Sheet.

6. Use the mouse pointer to observe the data at TP I for several complete transmission cycles. Describe what you observe in Question 2 on the Results Sheet.

7. Use the mouse pointer to observe several data output points for several complete transmission cycles. Describe what you observe in Question 3 on the Results Sheet.

DATA MULTIPLEXING

A Hands-On Project

This project uses an eight-channel CMOS analog multiplexer/demultiplexer circuit to demonstrate the operation of time-division multiplexing. You will:

• Construct the circuit.
• Apply various waveforms to the inputs and note the changes in the multiplexed output.

Preparation

1. Read Frenzel, *Principles of Electronic Communication Systems*, Section 10-3.

2. Complete the work for Prep Project 57.

Parts and Materials

2	Resistors, 15 kΩ
4	Resistors, 22 kΩ
1	IC, 4051 eight-channel analog multiplexer
2	SPST switches

The switches may be replaced with jumper wires.

Equipment

1	Function generator
1	Dual-trace oscilloscope
1	Power supply

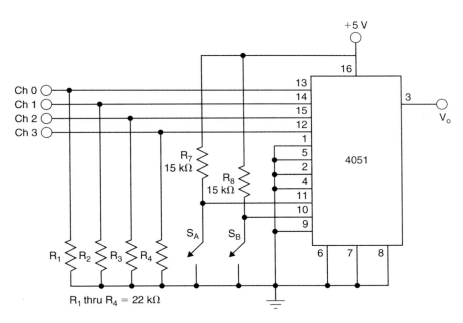

Figure 58-1

LAB PROCEDURE

1. Construct the circuit shown in Figure 58-1. Connect one channel of the oscilloscope to monitor the output of the circuit at V_{out}.

2. Adjust the function generator for a 100-Hz 6-V_{p-p}, sinusoidal waveform. Connect the function generator to the Ch 0 input of the circuit. Close switches SA and SB. This should send the signal at Ch 0 to the output of the multiplexer. Note the output of the circuit at V_a.

3. Connect the function generator to the Ch 3 input, and open both of the select switches. This action should send the signal at Ch 3 to the output of the multiplexer. Note the output of the circuit at V_{out}.

4. Set switches SA and SB to the four combinations of positions shown in Table 58-1. (For this circuit, closing a switch sets the logic-0 level, and opening the switch creates the logic-1 level.) In each case, connect the signal generator to each of the channels.

Name _____ Date _____

PROJECT 58

RESULTS SHEET

Table 58-1

Select Data		Data Input Channel at v_o
S_B	S_A	
0	0	
0	1	
1	0	
1	1	

Questions

1. Is the information being multiplexed in this project considered to be analog or digital data?

2. Is the output of this circuit considered to be analog or digital data?

Critical Thinking for Project 58

1. How many inputs are available to this circuit if *SB* is shorted to ground? Which inputs would be available?

2. How many select inputs would be required for an eight-channel TDM? Sixteen-channel TDM?

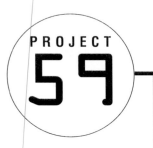
DATA DEMULTIPLEXING

A multiSIM Project

This project simulates the action of a digital data demultiplexer. You will:

- Observe the operation of the circuit with default settings.
- Note the effects of changing the scan frequency.

Preparation

Read Frenzel, *Principles of Electronic Communication Systems*, Section 10-3.

Setup Procedure

1. Start multiSIM on your computer.

2. Make sure that your communication lab CD-ROM is in the computer's CD drive.

3. Open the **multiSIM** directory on the CD-ROM.

4. Select **Project_59.msm**.

5. Look for a worksheet diagram that is similar to the one shown in Figure 59-1.

LAB PROCEDURE

This project is built around a three-line to eight-line digital demultiplexer. Three-bit binary (octal) codes at inputs A, B, and C determine which one of the eight outputs (D_0 through D_7) is energized. When one of the outputs is energized in this way, the logic level at input G_1 is directed to that output. In this project, G_1 is fixed at a high logic-1 level. So all outputs are low (logic-0) except for the one that is energized.

The data select inputs (A, B, and C) are operated here by an octal counter which generates binary values 000 through 111. This being the case, the outputs of the demultiplexer are energized in sequence—from D_0 through D_7. When you start the simulation, you will verify this fact by noting the action of the display lamps.

1. Start the simulation and observe the operation of the circuit.

2. Noting that the default frequency of V_1 is 1.5 kHz, calculate the period (T) of each clock pulse.

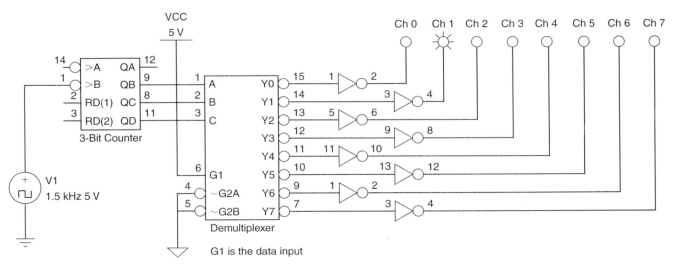

Figure 59-1

3. Calculate the amount of time required for completing one full demux cycle (T_D). *Hint:* Think about the modulus of the counter.

4. Stop the simulation and double-click the clock device (V_1) to show the Clock Source dialog box. Change the frequency to 3 kHz, and close the dialog box.

5. With the clock frequency now set to 3 kHz, calculate the period of each clock pulse and the time required for completing one full demux cycle.

RESULTS SHEET

STEP 2

  $f = 1.5$ kHz   T = _____

STEP 3

  T_D = _____

STEP 5

  $f = 3$ kHz   T = _____ T_D = _____

Questions

1. Knowing the clock frequency of V_1, how can you determine the length of time that each output is enabled?

2. How would the operation of this simulation be different if you switched the data input (G_1) from a fixed logic-1 level to a fixed logic-0 level?

Critical Thinking for Project 59

1. What is the main purpose of the logic inverters that are connected between the demultiplexer and display lamps?

2. Explain how time-domain demultiplexing is different in principle from frequency-domain demultiplexing.

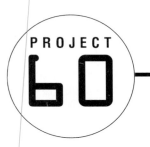

PROJECT 60

ANALOG MULTIPLEXER

A multiSIM Project

This project simulates the operation of a two-line to one-line analog multiplexer circuit. You will:

- Observe the basic operation of an analog multiplexer.
- Analyze the operating principles of a simple multiplexer.
- Calculate cycle rates.

Preparation

Read Frenzel, *Principles of Electronic Communication Systems*, Section 10-3.

Setup Procedure

1. Start multiSIM on your computer.

2. Make sure that your communication lab CD-ROM is in the computer's CD drive.

3. Open the **multiSIM** directory on the CD-ROM.

4. Select **Project_60.msm.**

5. Look for a worksheet diagram that is similar to the one shown in Figure 60-1.

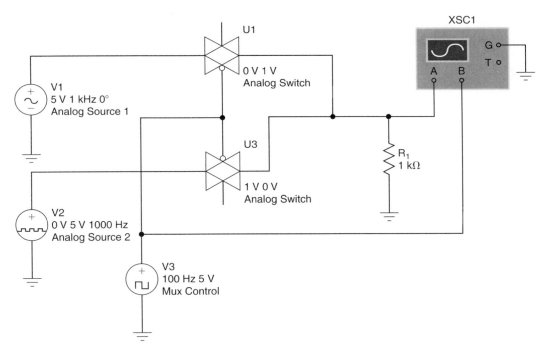

Figure 60-1

LAB PROCEDURE

PART 1 OBSERVING THE OPERATION

The purpose of this part of the project is simply to observe the operation of a 2:1 analog multiplexer.

1. Expand the oscilloscope instrument and start the simulation.

2. Make a sketch of the oscilloscope traces in Figure 60-2.

   Note that the output is a sine waveform when the Mux Control input is at its logic-1 level. The output is then a rectangular waveform when the Mux Control input is at its logic-0 level.

3. Use the oscilloscope as a measuring instrument to determine the values listed in Table 60-1.

PART 2 ANALYZING THE OPERATION

This circuit uses a pair of analog switches to direct source V_1 or source V_2 to the output resistor R_1. Source V_3 determines which analog switch is turned on at any given moment and, thus, which input signal is seen at the output.

1. Double-click analog switch U_1 to see the parameter box. Then select the Value tab to view the default values for the switch. Record the values in the appropriate spaces in Table 60-2.

2. Repeat the previous step for analog switch U_2.

3. Answer the questions for Step 3 in Part 2 on the Results Sheet.

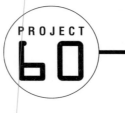

PROJECT
60

RESULTS SHEET

PART 1 OBSERVING THE OPERATION

STEP 2

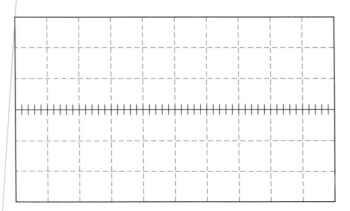

Figure 60-2

STEP 3

Table 60-1

Signal	Observed Value
Mux Control voltage	
Mux Control frequency	
Output sine wave amplitude (p–p)	
Output sine wave frequency	
Output rectangular wave amplitude	
Output rectangular wave frequency	

PART 2 ANALYZING THE OPERATION

Table 60-2

Analog Switch Parameter	Default Value
U1 COFF	
U1 CON	
U1 Roff	
U1 Ron	
U2 COFF	
U2 CON	
U2 Roff	
U2 Ron	

Questions

1. Referring to the information you recorded in Table 60-2, why is U_1 ON when the control voltage is equal or greater than $+1$ V, and why is U_2 ON when the control voltage drops to 0 V?

2. What would be the simplest way to go about changing this simulation so that U_1 is ON at -1 V and U_2 is ON at $+1$ V? (Try your theory if time permits.)

Critical Thinking for Project 60

1. Name at least one very significant difference between an analog multiplexer and a digital multiplexer. Name at least one significant similarity.

2. List the changes that would be necessary in order to expand this 2:1 multiplexer to a 4:1 version.

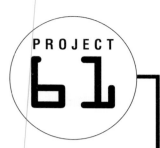

PROJECT 61

RF SWITCH

A commSIM Project

This project simulates the action of a 3:1 RF multiplexer followed by a comparable 1:3 RF demultiplexer. Although the frequencies for the simulation are in the range of 1 Hz to 10 Hz, they can represent actual frequencies of 1 GHz to 10 GHz. In this project you will:

- Observe the operation of a three-channel RF multiplexer/demultiplexer.
- Consider noise and signal losses in RF switches.

Preparation

Read Frenzel, *Principles of Electronic Communication Systems*, Section 10-3.

Setup Procedure

1. Start commSIM on your computer.

2. Make sure that your communication lab CD-ROM is in the computer's CD drive.

3. Open the **commSIM** directory on the CD-ROM.

4. Select **Project_61.vsm.**

5. Look for a worksheet diagram that is similar to the one shown in Figure 61-1.

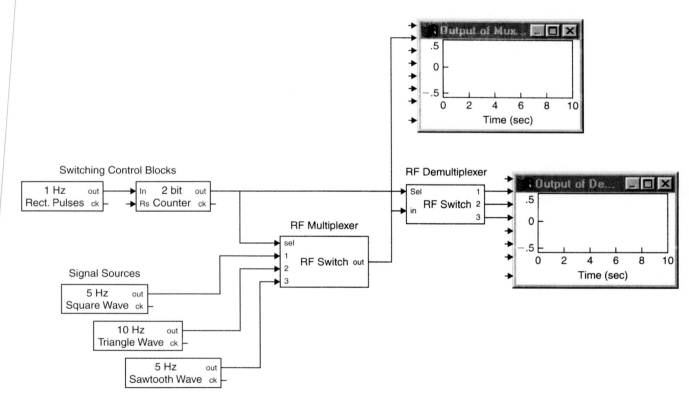

Figure 61-1

LAB PROCEDURE

1. Start the simulation with the default values.

2. Expand the plotter that is connected to the output of the multiplexer. Sketch the waveform in the graph provided as Figure 61-2.

3. Double-click each of the three signal sources to view their properties dialog boxes. Verify the default settings shown in Table 61-1.

4. Expand the plotter that is connected to the output of the RF demultiplexer. Sketch the waveform in Figure 61-3. Identify each RF channel with a number and an arrow pointing to a corresponding part of the signal.

Table 61-1

Signal Source	Shape	Frequency	Amplitude
1	Square	5 Hz	1 V
2	Triangle	10 Hz	1 V
3	Sawtooth	5 Hz	1 V

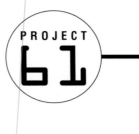

Name _____ Date _____

RESULTS SHEET

PROJECT 61

STEP 2

Figure 61-2

STEP 4

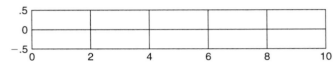

Figure 61-3

Questions

1. Why do the three signals at the output of the RF multiplexer appear in a stair-step configuration?

2. What is the modulus of the counter? The cycle frequency of the counter?

Critical Thinking for Project 61

1. Research and describe the difference between switch loss and isolation in an RF switch. See the Switch Properties dialog box for the RF Switch blocks in this commSIM project.

2. Explain why this simulation is an example of time-division multiplexing/demultiplexing.

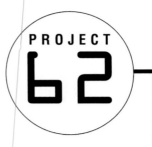

PROJECT 62

PARALLEL/SERIAL CONVERSION

An Extended Project

This project simulates the action of an 8-bit parallel-to-serial converter. You will:

- Enter 8-bit data into a parallel register.
- Shift the data out of the register serially.

Preparation

Read Frenzel, *Principles of Electronic Communication Systems*, Section 11-2.

Setup Procedure

1. Select **Extended Projects** from the **Projects** menu.

2. Select **Project 62   Serial/Parallel Conversion.**

LAB PROCEDURE

1. Enter binary 11111111 on the parallel input switches, and then press the load button to load the data into the register. Verify that the data is loaded, noting the status of the data lamps.

2. Repeatedly press the shift button, and notice how the bits are shifted from left to right.

3. Press the load button to load the same data, and then count the number of shift pulses required to send all 8 bits through the serial out port. Record the value on the Results Sheet.

4. Load 10000000 into the register and repeatedly click the shift button until the data is completely shifted out.

5. Load 00000001 into the register and repeatedly click the shift button until the data is completely shifted out. Record the number of shift pulses required to send all the data through the serial out port.

PROJECT
62

RESULTS SHEET

STEP 3

Number of shift pulses required = _____

STEP 5

Number of shift pulses required = _____

Questions

1. What is the hexadecimal value of the data entered in Step 1? What is the decimal value?

2. How many serial shift pulses are required to send 00000000?

Critical Thinking for Project 62

1. How many consecutive shift pulses would be required to pass 4-bit data through the serial out port? 16-bit data?

2. If the shift pulses occur at the rate of 320 kHz, how long does it take to send one complete word of 8-bit data?

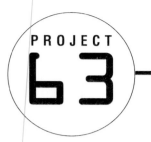

PROJECT 63 — MODEMS

An Extended Project

This project simulates the operation of modems that use quadrature amplitude modulation (QAM). You will:

- Select input data for a QAM and determine the output phase and amplitude of each input.
- Predict from your observations the output phase and amplitude for a given set of binary input data.
- Determine the binary input data required for producing a given QAM phase and amplitude.

Preparation

Read Frenzel, *Principles of Electronic Communication Systems*, Section 11-2.

Setup Procedure

1. Select **Extended Projects** from the **Projects** menu.

2. Select **Project 63   Modems.**

LAB PROCEDURE

The simulated modem device used in this project lets you generate 3- or 4-bit binary values that are immediately translated into a QAM signal. The signal is presented on a special oscilloscope display that directly indicates the signal's amplitude and phase angle.

For experimental purposes, you can select the operating mode for generating the binary data that is fed to the modem. Clicking the mode button cycles the modem through its manual count-up, count-down, and random modes of operation. When you are using one of the counting modes or the random mode, you can also select the speed at which the binary values change. Do this by clicking the rate push button: slow, medium, and fast.

When you are using the manual mode, click the data buttons to set up the binary value you want to transmit. Then press the load button to see the QAM conversion of that value.

PART 1   8-QAM MODULATOR

1. Set up each of the eight binary values listed in Table 63-1. Load each value and record the resulting QAM level on the Results Sheet. Record the phase angle in degrees, and the amplitude as high or low (inner circle or outer circle).

2. Set the modem for the count-up mode and the slow rate. See if you can verify findings you recorded in Table 63-1.

3. Set the modem for the random mode and fast speed. Bear in mind that this is still running far more slowly than the speeds found in actual QAM modems.

PART 2   16-QAM MODULATOR

1. Set up each of the 16 binary values listed in Table 63-2. Load each value and record the resulting QAM level on the Results Sheet. Record the phase angle in degrees, and the amplitude as high or low (inner circle or outer circle).

2. Set up each of the other operating modes for the modern. Observe the responses on the display.

PROJECT
63

RESULTS SHEET

PART 1   8-QAM MODULATOR

Table 63-1

D_3	D_2	D_1	D_0	Parity
0	0	0	0	
0	0	0	1	
0	0	1	0	
0	0	1	1	
0	1	0	0	
0	1	0	1	
0	1	1	0	
0	1	1	1	
1	0	0	0	
1	0	0	1	
1	0	1	0	
1	0	1	1	
1	1	0	0	
1	1	0	1	
1	1	1	0	
1	1	1	1	

Questions

1. Suppose a modem is receiving an 8-QAM signal of 135° high. What binary value does that signal represent?

2. Suppose D_1 of the modem input was always fixed at 0. Which angle-and-level combinations would be missing from the display?

PART 2   16-QAM MODULATOR

Table 63-2

D_3	D_2	D_1	D_0	Parity
0	0	0	0	
0	0	0	1	
0	0	1	0	
0	0	1	1	
0	1	0	0	
0	1	0	1	
0	1	1	0	
0	1	1	1	
1	0	0	0	
1	0	0	1	
1	0	1	0	
1	0	1	1	
1	1	0	0	
1	1	0	1	
1	1	1	0	
1	1	1	1	

Questions

1. Suppose a modem is receiving a 16-QAM signal of 135° high. What binary value does that signal represent?

2. Suppose a modem is receiving a 16-QAM signal of 247.5° high. What binary value does that signal represent?

Critical Thinking for Project 63

1. Explain why the display used in this project is sometimes called a *quadrature display*.

2. Suppose the modem's low voltage level was higher than it should be. How could this problem be detected on a quadrature display?

3. There is a practical reason for doubling the capacity of a QAM by doubling the number of phase angles rather than the number of different voltage levels. What is that reason?

BINARY PHASE-SHIFT KEYING

A multiSIM Project

This project simulates the operation of a BPSK modulator. You will:

- Observe the operation of the circuit with the default values.
- Vary the phase angle of the reference sine waveform and observe the results.
- Analyze the internal working of the BPSK simulator block.

Preparation

Read Frenzel, *Principles of Electronic Communication Systems*, Section 11-4.

Setup Procedure

1. Start multiSIM on your computer.

2. Make sure that your communication lab CD-ROM is in the computer's CD drive.

3. Open the **multiSIM** directory on the CD-ROM.

4. Select **Project_64.msm.**

5. Look for a worksheet diagram that is similar to the one shown in Figure 64-1.

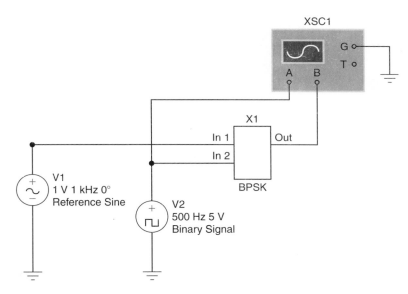

Figure 64-1

LAB PROCEDURE

1. Expand the oscilloscope instrument and start the simulation.

2. Sketch the waveforms on the graph in Figure 64-2.

   Carefully notice how the output waveform shows a 180° phase shift each time the binary signal changes state. Also notice that the shift takes place at (or very close to) 0° on the reference sine waveform.

3. Stop the simulation and double-click the Reference Sine symbol to bring up the AC Voltage dialog box. Notice that the default phase setting is 0°.

4. Set the phase value of the Reference Sine to 90°.

5. Close the dialog box and start the simulation.

6. Sketch the waveforms on the graph in Figure 64-2.

   Notice that the output waveform still shows a 180° phase shift, but at the 90° point on the reference sine wave rather than at the 0° point.

7. Double-click the BPSK box to bring up the Subcircuit index.

8. Click the Edit Subcircuit button to bring up the schematic diagram for the BPSK subcircuit. (Also see Figure 64-3 on the Results Sheet.)

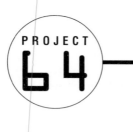

PROJECT
64

RESULTS SHEET

STEP 2

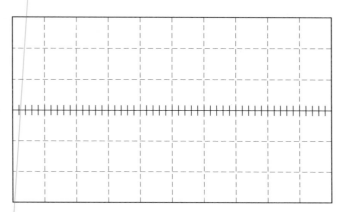

Figure 64-2

STEP 6

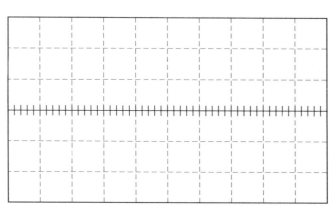

Figure 64-3

Questions

The following questions refer to Figure 64-4.

1. Which analog switch, U_2 or U_3, is switched on when the binary signal is at logic 0? At logic 1?

2. What is the voltage gain of amplifier U_1? Is this an inverting or noninverting amplifier?

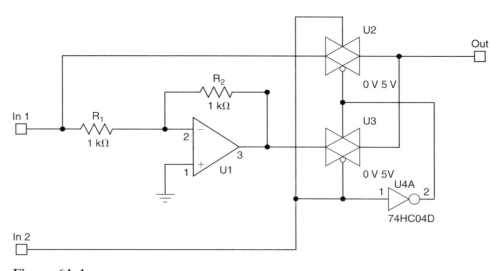

Figure 64-4

Critical Thinking for Project 64

1. Discuss why it is important that the sine and binary waveforms are of the same frequency and phase.

2. Explain why DPSK and QPSK techniques are preferred over BPSK.

PROJECT 65

PARITY GENERATOR/CHECKER

A Prep Project

This project is a computer simulation of odd and even parity generators and checkers. You will:

- Observe and test the operation of a 4-bit parity generator.
- Note the difference between odd and even parity.
- Observe and test the operation of a parity checker.
- Study the operation of a 4-bit parity generator/checker system.

Preparation

Read Frenzel, *Principles of Electronic Communication Systems*, Section 11-5.

Setup Procedure

1. Select **Prep Projects** from the **Projects** menu.

2. Select **Project 65   Parity Generator/Checker.**

LAB PROCEDURE

Throughout this project, a red lamp indicates a logic-1 level, and a black lamp indicates a logic-0 level. Also, a closed switch indicates a logic-1 level, and an open switch indicates a logic-0 level.

PART 1   PARITY GENERATOR

1. Click the O/E switch to set its closed status.

2. Set the input data switches to generate the 4-bit binary codes shown in Table 65-1. Record the corresponding status of the parity lamp.

3. Click the O/E switch to set its open status.

4. Set the input data switches to generate the 4-bit binary codes shown in Table 65-2. Record the corresponding status of the parity lamp.

PART 2   PARITY GENERATOR/ CHECKER

1. Click the O/E switch to set its closed status.

2. Set the input data switches to a series of different settings of your choice. Note that the received data follows your input data exactly. Describe on the Results Sheet the response of the parity alarm lamp for all of your data settings.

3. Click the O/E switch to set its open status.

4. Again, set the input data switches to a series of different settings of your choice. Note that the received data follows your input data exactly. Describe on the Results Sheet the response of the parity alarm lamp for all of your data settings.

Experimental Notes and Calculations

RESULTS SHEET

PART 1 PARITY GENERATOR

Table 65-1

D_3	D_2	D_1	D_0	Parity
0	0	0	0	
0	0	0	1	
0	0	1	0	
0	0	1	1	
0	1	0	0	
0	1	0	1	
0	1	1	0	
0	1	1	1	
1	0	0	0	
1	0	0	1	
1	0	1	0	
1	0	1	1	
1	1	0	0	
1	1	0	1	
1	1	1	0	
1	1	1	1	

Table 65-2

D_3	D_2	D_1	D_0	Parity
0	0	0	0	
0	0	0	1	
0	0	1	0	
0	0	1	1	
0	1	0	0	
0	1	0	1	
0	1	1	0	
0	1	1	1	
1	0	0	0	
1	0	0	1	
1	0	1	0	
1	0	1	1	
1	1	0	0	
1	1	0	1	
1	1	1	0	
1	1	1	1	

Questions

1. According to the data in Table 65-1, is this circuit operating as an odd-parity generator or an even-parity generator? Explain your answer.

2. According to the data in Table 65-2, is this circuit operating as an odd-parity generator or an even-parity generator? Explain your answer.

PART 2 PARITY GENERATOR/ CHECKER

STEP 2

The parity error lamp is

_____ ON continuously

_____ OFF continuously

_____ blinking ON and OFF

STEP 4

The parity error lamp is

_____ ON continuously

_____ OFF continuously

_____ blinking ON and OFF

Questions

1. What happens to the parity error lamp when the parity generator is operating as an even-parity generator?

2. What happens to the parity error lamp when the parity generator is operating as an odd-parity generator?

Critical Thinking for Project 65

1. Describe why it is important to match an even-parity generator with an even-parity receiver, and an odd-parity generator with an odd-parity receiver.

2. Describe what would be different about the operation of the parity error lamp if the D_0 bit at the receiver was always a 0, regardless of the status of the D_0 bit at the generator.

PARITY GENERATOR/CHECKER

A Hands-On Project

This project uses exclusive-OR logic gates to perform odd and even parity operations. The circuit in Part 1 of the project is a 4-bit parity generator that can operate as both an odd- and even-parity generator. This parity generator is then combined with a 4-bit parity checker in Part 2 of the project. You will:

- Construct the circuits.
- Apply binary inputs to a parity generator and observe the corresponding odd- and even-parity outputs.
- Determine the expected parity value of a binary number, and confirm your answer by actual experiment.
- Apply the output of the parity generator to a parity checker and observe correct and faulty data conditions.

Preparation

1. Read Frenzel, *Principles of Electronic Communication Systems*, Section 11-5.

2. Complete the work for Prep Project 65.

Parts and Materials

6	Resistors, 150 Ω
1	Resistor, 2.2 kΩ
2	ICs, 7486 TTL quad 2-input exclusive-OR gate
1	IC, 7493 TTL binary counter
6	Red LEDs
1	SPST switch

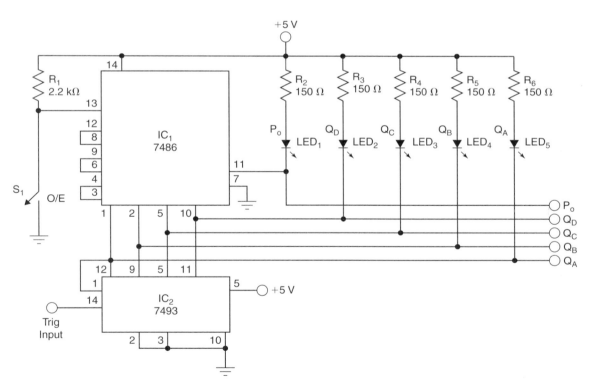

Figure 66-1

Equipment

1   DC power supply
1   Function generator

LAB PROCEDURE

PART 1   PARITY GENERATOR

1. Construct the circuit shown in Figure 66-1. Connect the TTL output of the function generator to the trigger input of the counter circuit.

2. Set the function generator to produce a 1-Hz waveform Apply power to the circuit and close the O/E switch.

    Carefully note the status of the lamps, and complete the data in Table 66-1. (Reduce the frequency of the function generator if the states are changing too quickly for you to note them accurately.)

3. Open the O/E switch and complete the data in Table 66-2.

PART 2   PARITY CHECKER

1. Construct the 4-bit parity checker circuit shown in Figure 66-2. Attach the data inputs to the outputs of the parity generator as indicated in Figure 66-3. Apply power and the function generator as before.

2. Close the O/E switch in the parity generator portion of the circuit and describe the condition of the parity error lamp as ON continuously, OFF continuously, or blinking ON and OFF.

3. Open the O/E switch in the parity generator portion of the circuit. Describe the condition of the parity error lamp as ON continuously, OFF continuously, or blinking ON and OFF.

4. Close the O/E switch. Remove the conductor that connects data line QB to IC3 in the parity checker. (This simulates one type of data corruption that occurs in the transmission of a data signal.) Describe the condition of the parity error lamp as ON continuously, OFF continuously, or blinking ON and OFF.

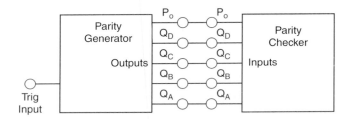

Figure 66-3

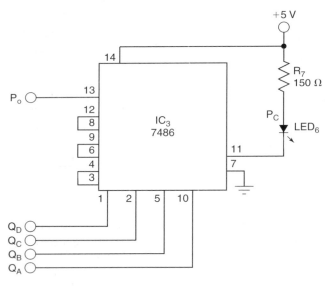

Figure 66-2

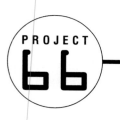

PROJECT 66

Name _____   Date _____

RESULTS SHEET

PART 1 PARITY GENERATOR

Table 66-1

D_3	D_2	D_1	D_0	Parity
0	0	0	0	
0	0	0	1	
0	0	1	0	
0	0	1	1	
0	1	0	0	
0	1	0	1	
0	1	1	0	
0	1	1	1	
1	0	0	0	
1	0	0	1	
1	0	1	0	
1	0	1	1	
1	1	0	0	
1	1	0	1	
1	1	1	0	
1	1	1	1	

Table 66-2

D_3	D_2	D_1	D_0	Parity
0	0	0	0	
0	0	0	1	
0	0	1	0	
0	0	1	1	
0	1	0	0	
0	1	0	1	
0	1	1	0	
0	1	1	1	
1	0	0	0	
1	0	0	1	
1	0	1	0	
1	0	1	1	
1	1	0	0	
1	1	0	1	
1	1	1	0	
1	1	1	1	

Questions

1. According to the data in Table 66-1, is the circuit operating as an odd-parity generator or an even-parity generator? Explain your answer.

2. According to the data in Table 66-2, is the circuit operating as an odd-parity generator or an even-parity generator? Explain your answer.

PART 2   PARITY CHECKER

STEP 2

The condition of the lamp is _____

STEP 3

The condition of the lamp is _____

STEP 4

The condition of the lamp is _____

STEP 5

The condition of the lamp is _____

Questions

1. How do you explain the results of Step 2?

2. How do you explain the results of Step 3?

3. How do you explain the results of Step 4?

Critical Thinking for Project 66

1. Suppose one of the data lines between the parity generator and checker in this project is open, but this fault condition is not visible. Explain how you could use careful observation of the operation of this circuit to determine which line is open.

2. Consider what happens when an even number of data bits is corrupted at the same time. Is it possible that no parity error would occur? Explain your answer.

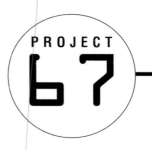

PINGING INTERNET SERVERS

A Hands-On Project

This project uses a pair of DOS commands, ping and tracert, to check the operation of your computer's Internet connection. You will:

- Ping a server to test the round-trip time of the data.
- Ping your own server to test its operation.
- Do a trace on the route a packet takes between your computer and another site.

Preparation

Read Frenzel, *Principles of Electronic Communication Systems*, Section 12-5.

Equipment

A PC running Windows and connected to the Internet.

LAB PROCEDURE

1. Open the MS-DOS Prompt window. (This is usually done by selecting the MS-DOS Prompt program from the Program menu.)

2. Obtain the URL (universal resource locator) of a working Web site. If your instructor does not provide one and you do not know any, use **glencoe.com** or **sweethaven.com.**

3. At the DOS prompt, type **ping *your_url*,** where *your_url* is the URL you decided to use. For example:

    ping glencoe.com

    or

    ping sweethaven.com

    Press the Return key to complete the command.

4. Note the response on the screen. On the Results Sheet, record:

    The URL you pinged
    The IP address of the server you pinged
    The number of replies from the server
    The number of bytes returned
    The round-trip time

5. Ping your own server with this DOS command: **ping localhost.**

6. Record the responses on the Results Sheet.

7. Run an analysis of the route that data packets take between your computer and a chosen URL: **tracert *your_url*,** where *your_url* is the URL you used in Step 3.

8. Note the number of different places your packets are visiting.

RESULTS SHEET

STEP 4

    URL: _____

    IP address: _____

    Number of replies: _____

    Number of bytes returned: _____

    Average round-trip time: _____

    The time to live (TTL) setting: _____

STEP 6

    URL: _____

    IP address: _____

Number of replies: _____

Number of bytes returned: _____

Average round-trip time: _____

The time to live (TTL) setting: _____

Questions

1. What happens if you ping a URL that does not exist or is off line?

2. Why does **tracert localhost** produce only one line of information?

Critical Thinking for Project 67

1. Discuss the value of the **ping** and **tracert** commands.

2. Most Web sites are identified by their URL. Describe how you can use the **ping** command to determine their corresponding IP address.

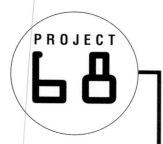

ANTENNA VOLTAGE AND CURRENT

PROJECT 68

A Prep Project

This project simulates the behavior of current and voltage for a dipole antenna that is tuned for half-wave resonance. You will:

- Adjust the carrier frequency applied to a half-wave dipole antenna for maximum current.
- Determine the physical length of a dipole, based on its half-wave resonant frequency.
- Measure and plot the voltage levels along a dipole antenna that is operating at half-wave resonance.

Preparation

Read Frenzel, *Principles of Electronic Communication Systems*, Section 14-1.

Setup Procedure

1. Select **Prep Projects** from the **Projects** menu.

2. Select **Project 68   Antenna Voltage and Current.**

LAB PROCEDURE

Referring to the block diagram on the screen, note that the RF generator supplies a signal to the dipole antenna.

PART 1   ANTENNA CURRENT

In this part of the project, you will use a simulated RF ammeter to monitor the amount of current flowing between the RF generator and the antenna. Note from the block diagram that the meter is connected in series with the transmission line.

1. Set the amplitude of the RF generator to 60 V.

2. Adjust the frequency of the RF generator to determine the half-wave resonant frequency of the antenna. This will be the frequency that causes the largest amount of current to flow to the antenna. Record this frequency on the Results Sheet.

PART 2   ANTENNA VOLTAGE

For this part of the project, you will use an RF voltmeter to determine the voltage along the length of a dipole antenna that is operating at half-wave resonance. The amplitude of the signal from the RF generator is adjustable, but the frequency is fixed at 500 MHz for this part of the work.

1. Move the mouse pointer to the antenna figure located on the right-hand side of the screen. Notice how the reading on the voltmeter changes as you move the pointer along the length of the antenna.

2. Set the mouse pointer to the positions indicated in Table 68-1 on the Results Sheet. Be sure you place the pointer directly on the antenna figure, and not on the labels that indicate the positions. Record the voltage levels you find at each of the locations specified in the table.

3. Plot the voltage levels from Table 68-1 in the space provided in Figure 68-1.

Experimental Notes and Calculations

PROJECT 68

RESULTS SHEET

PART 1 ANTENNA CURRENT

STEP 2

Frequency = _____

Questions

1. Based on your data of Step 2, what is the physical length of this half-wave dipole antenna?

2. If the length of this antenna were physically short-ened, would you have to increase or decrease the applied frequency in order to restore half-wave resonance?

PART 2 ANTENNA VOLTAGE

Table 68-1

Position	Voltage
0	
$\lambda/16$	
$\lambda/8$	
$3\lambda/16$	
$\lambda/4$	
$5\lambda/16$	
$3\lambda/8$	
$7\lambda/16$	
$\lambda/2$	

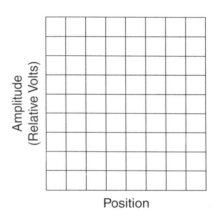

Figure 68-1

Questions

1. Why does this experiment show the same polarity of signal at both ends of the antenna when, in fact, they are of opposite polarity?

2. At what frequency would this antenna be a quarter-wave antenna?

Critical Thinking for Project 68

1. Explain why it is necessary to peak the antenna current (as in Part 1) before you can properly plot its half-wavelength electrical field (as in Part 2).

2. Explain the principle of antenna reciprocity. Suggest how it could apply to the antenna in this project.

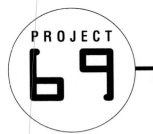

ANTENNA VOLTAGE AND CURRENT

A Hands-On Project

This project uses an RF signal generator and simple measuring instruments to determine the voltage standing wave ratio (SWR) of a transmission line and dipole antenna. You will:

- Construct the antenna.
- Tune the system for maximum current.
- Measure relative voltage levels along the length of the antenna.

Preparation

1. Read Frenzel, *Principles of Electronic Communication Systems*, Section 14-1.

2. Complete the work for Prep Project 68.

Components and Supplies

1	Capacitor, 100 pF
1	Diode, lN34 or equivalent
1	DC microammeter, 0–50 μA
1	Terminal board, 2-position
2	8-penny nails
	6 in of 10-lb nylon fishing line
	Wooden board, 1 × 4, 4½ ft long
	2 ft 2 in of 300-Ω twin lead transmission line
	4 ft 6 in of bare copper wire, about 18 gauge

Equipment

1	RF signal generator
1	Tape measure

LAB PROCEDURE

PART 1 ANTENNA ASSEMBLY

Construct the antenna assembly shown in Figure 69-1. Here are a few hints:

1. Form the two elements of the dipole antenna by cutting the copper wire exactly in half.

2. The critical dimension is the end-to-end length of the copper-wire dipole—it should be very close to the specified length of 4 ft 3 in.

3. The terminal board should be centered between the ends of the dipole elements.

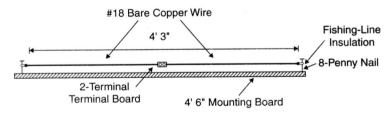

Figure 69-1

PART 2 ANTENNA CURRENT

The real purpose of this part of the project is to make the best possible match between the length of the dipole, the length and characteristic impedance of the transmission line, and the frequency of the RF generator.

1. Connect the transmission line, meter, diode, and RF generator, as shown in Figure 69-2. The meter assembly should be connected in series with the "ungrounded" lead from the RF generator.

2. Set the frequency of the RF generator to about 106 MHz. Adjust the amplitude of the RF signal to a point at which the deflection of the meter movement is obvious.

3. Adjust the frequency of the RF generator to peak the antenna current. If the meter pegs before the peak is reached, reduce the signal amplitude. Record the peak frequency on the Results Sheet.

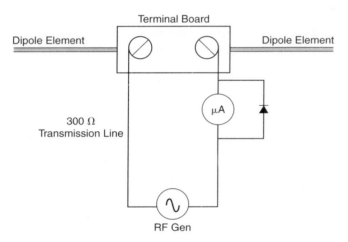

Figure 69-2

PART 3 ANTENNA VOLTAGE

Important: Part 2 must always be completed prior to starting this part. There must be no adjustments to the length of the dipole elements or changes in the frequency setting of the RF generator.

1. Remove the meter assembly from the circuit, and attach both conductors of the transmission line directly to the terminal board and dipole elements. Construct the RF "probe" assembly also, as shown in Figure 69-3.

2. Complete the ground connection to the probe assembly by holding the connection between your fingers. Slide the open end of the capacitor along the elements of the dipole antenna. Note the maximum and minimum responses. Increase the amplitude of the RF generator to get the largest possible amount of response for the maximum readings.

Remember: Do not change the frequency setting from the value established in Part 2 of this project.

3. Locate the probe at the outside tip of one of the dipoles. This will be your zero reference location. Record the meter reading in the *V* column of Table 69-1 on the Results Sheet.

4. Calculate the value of $\lambda/16$ in units of inches. Record your value in the *d* column of the row labeled $\lambda/16$ in Table 69-1. Locate that point on your antenna, and then take a meter reading at that point. Record your results in the *V* column of the $\lambda/16$ row.

5. Repeat the procedure of Step 4 for all of the locations (except $\lambda/4$) listed in Table 69-1.

6. Plot the results of your measurements in Figure 69-4.

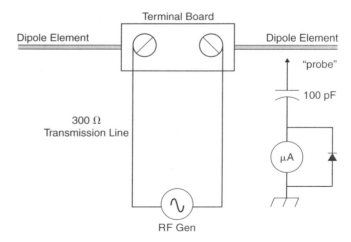

Figure 69-3

Name _____   Date _____

RESULTS SHEET

PART 2   ANTENNA CURRENT

STEP 3

Frequency at peak current = _____

Questions

1. At the frequency of Step 3, is this antenna operating as a full-, half-, or quarter-wave dipole?

2. Given the physical length of this dipole, what is its calculated resonant frequency?

3. How do you account for any difference between the results in Step 3 and your calculation in Question 2?

PART 3   ANTENNA VOLTAGE

Table 69-1

Position	Position d in inches	Voltage V relative
0		
λ/16		
λ/8		
3λ/16		
λ/4		
5λ/16		
3λ/8		
7λ/16		
λ/2		

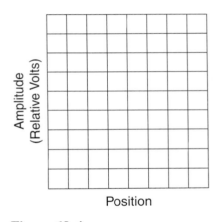

Figure 69-4

Questions

1. What is the purpose of Part 2 of this project?

2. Why is a reading for λ/4 omitted from Table 69-1?

Critical Thinking for Project 69

1. Explain why it is important to perform Part 2 of this project just prior to performing Part 3.

2. Explain the purpose of the diode that is connected across the terminals of the microammeter.

ANTENNA IMPEDANCE MATCHING

An Extended Project

This project simulates the use of a quarter-wave transformer to match the impedance between a transmission line and antenna. You will:

- Determine the amount of impedance mismatch between a transmission line and antenna.
- Observe the standing wave ratio (SWR) of a transmission line and antenna system.
- Calculate the transformer impedance required for reducing the SWR to 1.
- Adjust the separation of conductors in a quarter-wave transformer to reduce the SWR as close to 1 as possible.

Preparation

Read Frenzel, *Principles of Electronic Communication Systems*, Section 14-2.

Setup Procedure

1. Select **Extended Projects** from the **Projects** menu.

2. Select **Project 70   Antenna Impedance Matching** from the menu.

LAB PROCEDURE

This project provides the means for establishing a series of impedance matches and mismatches between a transmission line and antenna. This is done by clicking the buttons located beside the readouts for transmission line and antenna impedance. Digital readouts on the experimental unit also indicate the current values of the quarter-wave transformer and the resulting SWR.

The objective of the work in this project is to vary the impedance of the quarter-wave transformer by adjusting the spacing between the two simulated conductors. This adjustment is made by moving the mouse pointer to one of the elements of the transformer, pressing the left mouse button, and dragging the conductors closer or farther apart. This matching process is completed when you can adjust the transformer to a value of SWR as close to 1 as it can possibly be.

1. Set the line impedance to 50 Ω. Set the antenna impedance to the four values shown in Table 70-1. In each case, adjust the spacing between the conductors of the quarter-wave transformer to obtain the smallest possible SWR. Record the values for this SWR and matching impedance in the table.

2. Set the line impedance to 75 Ω. Repeat the procedure in Step 1, using Table 70-2.

3. Set the line impedance to 150 Ω. Repeat the procedure used in the previous steps. Use Table 70-3 on the Results Sheet.

4. Set the line impedance to 300 Ω. Repeat the procedure used in the previous steps. Use Table 70-4 on the Results Sheet.

RESULTS SHEET

Table 70-1

Antenna Impedance	Smallest SWR	Matching Impedance
50 Ω		
75 Ω		
150 Ω		
300 Ω		

Table 70-2

Antenna Impedance	Smallest SWR	Matching Impedance
50 Ω		
75 Ω		
150 Ω		
300 Ω		

Table 70-3

Antenna Impedance	Smallest SWR	Matching Impedance
50 Ω		
75 Ω		
150 Ω		
300 Ω		

Table 70-4

Antenna Impedance	Smallest SWR	Matching Impedance
50 Ω		
75 Ω		
150 Ω		
300 Ω		

Questions

1. What amount of matching impedance is required when the transmission line impedance already matches the antenna impedance?

2. What kind of impedance match or mismatch is necessary for achieving a value of SWR that is less than 1?

Critical Thinking for Project 70

1. Under certain conditions, the spacing between the conductors of the matching transformer will be less than the spacing of the conductors of the transmission line. Under the opposite set of conditions, the spacing is greater. Describe these conditions and explain why wider or narrower spacing is required.

2. This project deals only with the spacing of conductors in a quarter-wave transformer. Name two other factors that also affect the impedance.

3. Briefly describe why it is important to locate a quarter-wave transformer very close to the antenna.

PROJECT 71

MICROWAVE SYSTEMS

A Virtual Project

This project allows you to determine frequency and voltage levels in a simulated microwave transmitter. You will:

- Measure and record voltage and frequencies on an interactive block diagram of a microwave system.
- Determine the frequency multiplication and voltage gain of selected sections of the circuit.

Preparation

Read Frenzel, *Principles of Electronic Communication Systems*, Section 15-1.

Setup Procedure

1. Select **Virtual Projects** from the **Projects** menu.

2. Select **Project 71   Microwave Systems.**

LAB PROCEDURE

This project uses an interactive block diagram. Clicking a test point on the diagram simulates connecting a test probe to it. This probe is connected to both a frequency counter and an AC voltmeter.

1. Attach the test probe to each of the seven test points as listed in Table 71-1. Record the frequency and voltage amplitude at each point.

2. Answer all the questions listed on the Results Sheet.

PROJECT
71

RESULTS SHEET

Table 71-1

Test Point	Frequency	Amplitude
TP 1		
TP 2		
TP 3		
TP 4		

STEP 2

1. What is the operating frequency of the oscillator?

   _____

2. What is the frequency multiplication factor for

   amplifier 1? _____

3. What is the voltage gain, in dB, for amplifier 1?

   _____

4. What is the frequency multiplication factor for

   amplifier 2? _____

5. What is the voltage gain, in dB, for amplifier 2?

   _____

6. What is the frequency multiplication factor for

   amplifier 3? _____

7. What is the voltage gain, in dB, for amplifier 3?

   _____

8. What is the frequency multiplication factor for

   amplifier 4? _____

9. What is the voltage gain, in dB, for amplifier 4?

   _____

10. What is the overall frequency multiplication for

    amplifier 1 through amplifier 4? _____

11. What is the overall voltage gain, in dB, for

    amplifier 1 through amplifier 4? _____

Critical Thinking for Project 71

1. Explain how you know whether this block diagram represents an AM or an FM transmitter.

2. Cite the primary purpose of amplifier 1 through amplifier 4.

3. Explain how it is possible to see a positive voltage gain through the bandpass filter in this block diagram.

This project simulates the operation of a telephone pulse-tone generator. You will:

* Note the two frequencies used for each button on the standard telephone keypad.
* Listen to the frequencies individually.
* Listen to the frequencies as combined for each button on the keypad.

Preparation

Read Frenzel, *Principles of Electronic Communication Systems*, Section 17-1.

Setup Procedure

1. Select **Extended Projects** from the **Project** menu.

2. Select **Project 72   Pulse-Tone Dialer.**

LAB PROCEDURE

This work screen uses two devices: a typical telephone keypad and a form showing the touch-tone matrix. Experiment with the various tones and combinations of tones. If a telephone is available nearby, compare the touch tones from the computer with those from an actual telephone.

PROJECT
72

RESULTS SHEET

Critical Thinking for Project 72

1. Describe why a pair of tones, rather than just one tone, is transmitted for each key.

2. Explain why none of the basic frequencies of the touchtone system is a harmonic of any other of the basic frequencies.

FIBER-OPTIC COMMUNICATION

A Hands-On Project

This project uses a standard fiber optic demonstration kit. You will:

- Assemble the transmitter and receiver circuits.
- Prepare the fiber optic couplings.
- Connect the transmitter and receiver via the optical fiber.
- Compare the transmitted and received waveforms.

Preparation

Read Frenzel, *Principles of Electronic Communication Systems*, Sections 18-2, 18-3, and 18-4.

Components and Supplies

1 Fiber optic communication kit, such as Industrial Fiber Optics #IF-E22

Industrial Fiber Optics
627 South 48th Street, Suite 100
Tempe, AZ 85281
www.i-fiberoptics.com

Equipment

1   DC power supply
1   Function generator
1   Dual-trace oscilloscope

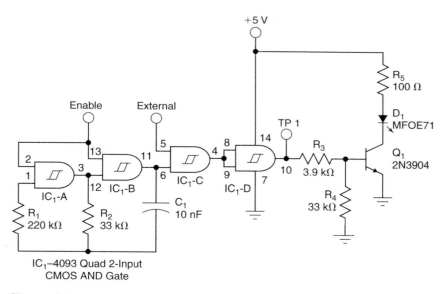

Figure 73-1

LAB PROCEDURE

PART 1 IR TRANSMITTER AND RECEIVER

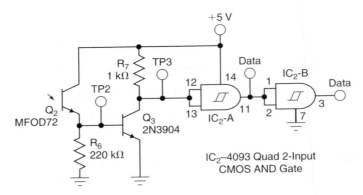

Figure 73-2

1. Assemble the transmitter and receiver circuits shown in Figures 73-1 and 73-2. If you are using the recommended kit, follow the instructions in the manual for mounting components onto the printed circuit boards.

2. Connect the transmitter circuit to the +5 V DC power source and apply power. Connect a jumper from the EN terminal to the negative (or ground) side of the power supply, and the EXT terminal to the positive side. Carefully observe the output of the LED device. You should be able to see a faint red glow, which indicates the LED is turned on. When you switch the EXT jumper to the negative side of the power supply, the LED should turn off.

3. Remove the jumper from the EXT terminal, and replace it with the TTL output of the function generator. Set the frequency of the function generator for a 2 or 3 Hz rectangular waveform, and closely observe the light from the LED. It should be blinking at the rate of the signal from the function generator.

4. Connect the receiver circuit to the +5 V DC source. Connect CHB of the oscilloscope to the DATA output terminal of the receiver. Your setup should resemble the one shown in Figure 73-3. Arrange the two circuit boards so that the LED and phototransistor are facing one another, about a half-inch apart. Adjust the trace on the oscilloscope to see the waveform at the DATA output terminal. Sketch the waveform in Figure 74-4 on the Results Sheet.

5. Move the transmitter farther away from the receiver, but try to maintain the communication link as indicated by the waveform at the DATA terminal of the receiver. Note how far the transmitter and receiver can be separated before the DATA signal drops off −3 dB or more. Record the approximate distance on the Results Sheet.

6. Move the transmitter to about 1 inch from the receiver and make sure you have a good signal at the DATA terminal. Slide a piece of paper between the LED and phototransistor. Record the result on the Results Sheet.

PART 2 FIBER-OPTIC LINK

1. Prepare the fiber-optic cable and connectors as described in the instruction manual supplied with the fiber-optic kit.

2. Fasten the cable between the LED on the transmitter circuit and phototransistor on the receiver.

3. With the output of the function generator set for a 2 or 3 Hz rectangular waveform, adjust the oscilloscope to view the receiver waveform at the data terminal on the receiver. Record this waveform in Figure 73-4 on the Results Sheet.

4. Move the transmitter and receiver as far apart as possible, and rotate one of the circuit boards so that the LED and phototransistor are no longer aligned. Describe the effect these changes have upon the quality of the signal at the data terminal.

PROJECT 73

RESULTS SHEET

PART 1 IR TRANSMITTER AND RECEIVER

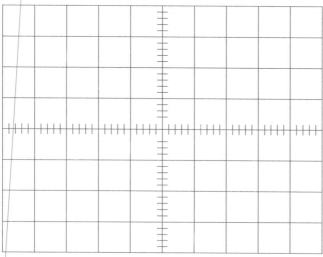

Figure 73-3

STEP 5

Maximum effective operating distance = _____

STEP 6

The effect of blocking the signal is

_____

Questions

1. Is the modulator in the transmitter circuit an analog or digital type?

2. Is the light-generating device in the transmitter circuit a true laser source?

PART 2 FIBER-OPTIC LINK

Figure 73-4

STEP 4

The effect of moving around the transmitter and receiver is

_____

Questions

1. What is the advantage of using a fiber-optic link as opposed to transmission through air?

2. What type of device is used as the light sensor in the receiver circuit?

315

Critical Thinking for Project 73

1. Explain the principle of intensity modulation and compare it with RF amplitude modulation.

2. Explain the advantages of using a laser diode as a light source as opposed to a light-emitting diode.

PROJECT 74 — TELEVISION SYSTEM

An Extended Project

This project gives you a chance to explore and troubleshoot an interactive block diagram of a modern color TV receiver. Each block represents an IC device or a group of closely related electronic components. The lines connecting the blocks represent signal paths and electrical connections. Horizontal and vertical scroll bars on the presentation let you move the diagram around so that you can access any portion of it. You will:

- Explore the purpose of each block in the interactive diagram.
- Examine the waveforms commonly found throughout a TV receiver system.
- Troubleshoot and suggest the type of repair that is required.

Preparation

Read Frenzel, *Principles of Electronic Communication Systems*, Sections 19-2.

Setup Procedure

1. Select **Extended Projects** from the **Projects** menu.

2. Select **Project 74   Television System**.

LAB PROCEDURE

You will use the mouse to point and click the various blocks and lines to determine some essential technical details about the various units of a TV receiver and the electrical signals that flow between them. When Interactive TV Lab is run in its Tutorial mode, the information and technical details represent a TV receiver that is operating properly. In the Trouble mode, however, the technical information clearly indicates the nature of a trouble in the receiver.

PART 1   EXPLORING THE INTERACTIVE BLOCK DIAGRAM

The objective of this part of the project is to explore all of the basic TV blocks and signal lines.

1. Make sure that the program is in the Tutorial mode.

2. Click each of the blocks on the diagram and read the corresponding descriptions in the Description box.

3. Click each of the lines connecting the blocks and note the corresponding description, voltage levels, and oscilloscope displays.

PART 2 TROUBLESHOOTING FAULTS IN THE SYSTEM

This part of the project uses the troubleshooting features that are available with this interactive block diagram.

1. Click the Trouble button to change to the troubleshooting mode of operation.

2. Note the symptoms of a trouble in the Symptom(s) box.

3. When you believe that you have determined the trouble, click the Fix It button. This brings up the TV Fix-It Form.

4. Select your remedy from the list and click the Okay button.

5. The program will consider your remedy and let you know whether it cures the trouble.

6. Repeat Steps 1 through 6 as long as lab time permits or once you are no longer seeing new troubles.

TELEVISION REMOTE CONTROLS

PROJECT 75

A Prep Project

The simulated instruments and circuits in this project include a commercial TV remote control unit, an infrared receiver, and an oscilloscope. In this project your will:

• Observe the sequence of pulses generated by a commercial TV remote control.
• Use an IR receiver and oscilloscope to determine the pulse code pattern of the TV remote control.

Preparation

Read Frenzel, *Principles of Electronic Communication Systems*, Section 19-2 and 20-6.

Setup Procedure

1. Select **Prep Projects** from the **Projects** menu.

2. Select **Project 75   Television Remote Controls.**

LAB PROCEDURE

1. Click the buttons on the TV remote unit at random. Note the type of response appearing on the oscilloscope display. Determine by experiment and observation the number of bits this unit is transmitting. Record your answer on the Results Sheet.

2. Click the buttons in the sequence outlined in Table 75-1. Record the status of each pulse position as *hi* (pulse on) or *lo* (pulse off).

3. Sketch the pulse train for Key 9 in Figure 75-1.

Name _____ Date _____

RESULTS SHEET

STEP 1

Number of bit positions used = _____

STEP 2

Table 75-1

Key Pressed	Pulse 1	Pulse 2	Pulse 3	Pulse 4	Pulse 5	Pulse 6	Pulse 7	Pulse 8
0								
1								
2								
3								
4								
5								
6								
7								
8								
9								
VOL UP								
VOL DN								
CH UP								
CH DN								

STEP 3

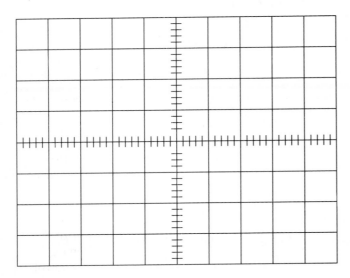

Figure 75-1

Questions

1. If the pulses are 1 ms apart, what is the total period of the longest pulse sequence?

2. If the sweep rate of the oscilloscope is 1 ms/div, what is the width of a hi pulse?

Critical Thinking for Project 75

1. Explain why it is important that the first pulse in each transmission sequence is always a hi pulse.

2. Determine the number of pulse positions required for a TV remote that has 24 different buttons.

3. Earlier TV remote control sometimes used ultrasound rather than IR transmission. What is a major disadvantage of the ultrasound medium?

TELEVISION REMOTE CONTROLS

PROJECT 76

A Hands-On Project

This project provides the opportunity to observe and decode the infrared signal produced by a conventional TV remote control. You will:

- Set up the equipment required for observing the IR signal.
- Determine the number of pulses in the pulse-train sequence.
- Determine the coding that is used for keypad numerals 0 through 9.

Preparation

1. Read Frenzel, *Principles of Electronic Communication Systems*, Sections 19-2 and 20-6.

2. Complete the work for Prep Project 76.

Components and Supplies

1 IR receiver (from Project 75)
1 TV remote control unit

Equipment

1 Oscilloscope

LAB PROCEDURE

1. Obtain the IR receiver circuit constructed in Project 76 and connect the oscilloscope to monitor the data output of the receiver.

2. Bring the IR emitter window of the TV remote control within an inch of the phototransistor of the IR receiver. Click the buttons on the TV remote at random, adjusting the sweep and amplitude controls on the oscilloscope to show the IR response of the receiver. Set the oscilloscope for triggering. Adjust the trigger slope to show a trace only when one of the buttons on the remote is pressed.

3. Note the type of response appearing on the oscilloscope display. Determine by experiment and observation the number of bits this unit is transmitting. Record your findings on the Results Sheet.

4. Sketch the pulse train for the signal that is transmitted while pressing Key 9. Use the graph provided in Figure 76-1.

5. Press the buttons in the sequence outlined in Table 76-2. Record the status of each pulse position as hi (pulse on) or lo (pulse off).

PROJECT 76

RESULTS SHEET

STEP 3

   Number of pulses per train = _____

STEP 4

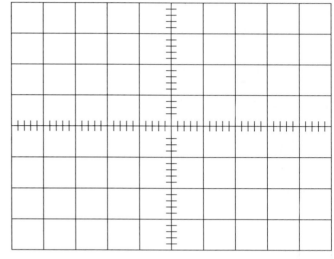

Figure 76-1

STEP 5

Table 76-1

Key Pressed	Pulse 1	Pulse 2	Pulse 3	Pulse 4	Pulse 5	Pulse 6	Pulse 7	Pulse 8
0								
1								
2								
3								
4								
5								
6								
7								
8								
VOL UP								
VOL DN								
CH UP								
CH DN								

Questions

1. What is the time allocated for each pulse? For the entire pulse train?

2. Does your unit automatically repeat the transmission when you hold down a key?

Critical Thinking for Project 76

1. Research and explain the nonreturn to zero (NRZ) series code that is commonly used with TV remote controls.

2. Suggest some practical ways to multiply the output power of a TV remote control.

PROJECT 77

LISSAJOUS PATTERNS

A multiSIM Project

Preparation

Read Frenzel, *Principles of Electronic Communication Systems*, Section 21-2.

Setup Procedure

1. Start multiSIM on your computer.

2. Make sure that your communication lab CD-ROM is in the computer's CD drive.

3. Open the **multiSIM** directory on the CD-ROM.

4. Select **Project_77.msm.**

5. Look for a multiSIM diagram that is similar to the one shown in Figure 77-1.

LAB PROCEDURE

PART 1 EQUAL FREQUENCIES AND PHASE ANGLES

Table 77-1 shows the default settings for this project—equal frequencies and phase angles.

1. Start the simulation and expand oscilloscope XSC1. You should see the two waveforms presented as a function of time. Notice that they have the same amplitude, frequency, and phase.

2. Expand oscilloscope XSC2 to see the Lissajous pattern. Sketch this figure in the space provided in Table 77-1.

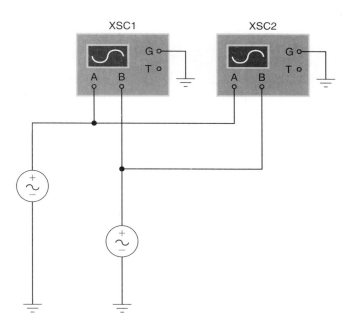

Figure 77-1

PART 2 EQUAL FREQUENCIES, DIFFERENT PHASE ANGLES

In this part of the project, you will vary the phase angle of the waveforms and compare their Lissajous patterns. For each of the settings listed in Table 77-2:

1. Make sure that the simulation is turned off.

2. Set up the prescribed values for the input waveforms.

3. Start the simulation.

4. Observe the more "traditional," time-domain versions of the waveforms on XCS1.

5. Open XCS2 and sketch the corresponding Lissajous pattern in the space provided in Table 77-2.

PART 3 DIFFERENT FREQUENCIES, EQUAL PHASES

Now you will compare Lissajous patterns that occur when the waveforms have different frequencies. The following instructions refer to Table 77-3:

1. Turn off the simulation.

2. Set up the frequencies specified in the table. (Make sure that the phase angles are 0 in all instances.)

3. Start the simulation.

4. Observe the more "traditional," time-domain versions of the waveforms on XCS1.

5. Open XCS2 and sketch the corresponding Lissajous pattern in the space provided in the table.

PROJECT 77

RESULTS SHEET

PART 1 EQUAL FREQUENCIES AND PHASE ANGLES

PART 2 EQUAL FREQUENCIES, DIFFERENT PHASE ANGLES

Table 77-1

Inputs	Lissajous Pattern

Input A (V1)
   Frequency = 1 kHz
   Phase = 0°
Input B (V2)
   Frequency = 1 kHz
   Phase = 0°	

Questions

1. What are the essential features of the Lissajous pattern for two waveforms that have the same frequency and phase angle? (*Hint:* Consider the shape of the figure and the angle it makes with the horizontal axis of the display).

2. What difference, if any, should you see in the Lissajous pattern if you switch the A and B inputs?

Table 77-2

Inputs	Lissajous Pattern

Input A (V1)
   Frequency = 1 kHz
   Phase = 45°
Input B (V2)
   Frequency = 1 kHz
   Phase = 0°	

Input A (V1)
   Frequency = 1 kHz
   Phase = 90°
Input B (V2)
   Frequency = 1 kHz
   Phase = 0°	

Input A (V1)
   Frequency = 1 kHz
   Phase = 180°
Input B (V2)
   Frequency = 1 kHz
   Phase = 0°	

Input A (V1)
   Frequency = 1 kHz
   Phase = 0°
Input B (V2)
   Frequency = 1 kHz
   Phase = 90°	

Input A (V1)
   Frequency = 1 kHz
   Phase = 0°
Input B (V2)
   Frequency = 1 kHz
   Phase = 45°	

Questions

1. How would you verbally describe the Lissajous pattern that occurs when the input waveforms have equal frequency but are 45° out of phase?

2. What is the Lissajous pattern that you find when the inputs are of equal frequency but 90° out of phase?

3. How does the Lissajous pattern for a 180° phase angle compare with one for a 0° phase angle?

PART 3   DIFFERENT FREQUENCIES, EQUAL PHASES

Table 77-3

Inputs	Lissajous Pattern

Input A (V1)
    Frequency = 1 kHz
    Phase = 0°
Input B (V2)
    Frequency = 2 kHz
    Phase = 0°	

Input A (V1)
    Frequency = 1 kHz
    Phase = 0°
Input B (V2)
    Frequency = 4 kHz
    Phase = 0°	

Input A (V1)
    Frequency = 4 kHz
    Phase = 0°
Input B (V2)
    Frequency = 1 kHz
    Phase = 0°	

Input A (V1)
    Frequency = 2 kHz
    Phase = 0°
Input B (V2)
    Frequency = 1 kHz
    Phase = 0°	

Questions

1. Which frequency multiple produces a Lissajous pattern commonly known as the *bow-tie figure?*

2. You have seen the number of peaks, or crowns, that occur when the frequencies are 2 × and 4 ×. How many peaks will the Lissajous patterns have for frequency multiples of 6 × and 8 ×?

Critical Thinking for Project 77

1. Name at least one practical application for Lissajous patterns in communication electronics testing.

2. Most real-world oscilloscopes have inputs labeled X and Y. How do these designations relate to inputs A and B for this project?

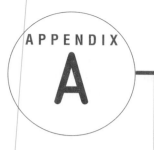

COMPOSITE PARTS AND EQUIPMENT LISTS

Resistors

All fixed resistors are 1/4 W, 5%

1	Resistor, 100 Ω
1	Resistor, 120 Ω
1	Resistor, 150 Ω
1	Resistor, 180 Ω
1	Resistor, 300 Ω
2	Resistor, 330 Ω
1	Resistor, 470 Ω
1	Resistor, 820 Ω
2	Resistor, 1 kΩ
2	Resistor, 1.5 kΩ
2	Resistor, 2.2 kΩ
1	Resistor, 2.7 kΩ
1	Resistor, 3.3 kΩ
1	Resistor, 4.7 kΩ
5	Resistor, 10 kΩ
4	Resistor, 20 kΩ
1	Resistor, 22 kΩ
1	Resistor, 27 kΩ
1	Resistor, 47 kΩ
1	Resistor, 100 kΩ
1	Resistor, 330 kΩ
1	Potentiometer, 10 kΩ

Capacitors

2	Capacitor, 100 pF
2	Capacitor, 1 nF
2	Capacitor, 10 nF
1	Capacitor, 68 nF
2	Capacitor, 100 nF
1	Capacitor, 1 μF
1	Capacitor, 10 μF
1	Trimmer capacitor, 20–90 pF

Inductors

1	Inductor, 1 mH

Discrete Semiconductors

1	Diode, 1N34 or equivalent
1	NPN transistor, 2N3904 or 2N2222
1	JFET, 2N5457

IC Devices

1	IC, LM386 audio power amplifier
2	IC, 555 timer
1	IC, 565 phase-locked loop
1	IC, 566 voltage-controlled oscillator
1	IC, 741 op-amp
1	IC, ADC0804 8-bit A/D converter
1	IC, 3080 operational transconductance amplifier
1	IC, 7404 TTL hex inverter
1	IC, 7493 binary counter
1	IC, 4051 8-channel analog multiplexer

Miscellaneous Electronic Parts

1	455 kHz ceramic filter
1	Crystal, 5.00 MHz
1	Crystal, 3.579 MHz
1	8 Ω permanent-magnet speaker or earphone
1	Fiber optic communication kit, such as Industrial Fiber Optics #IF-E22B

Miscellaneous Supplies

1	Terminal board, 2-position
10 ft.	Copper wire, about 18 gauge
2 ft. 2 in.	300 Ω twin lead transmission line
2	8-penny nail
4 ft. 6 in.	bare copper wire, about 18 gauge
4	SPDT switch
6 in.	10 lb nylon fishing line
	Wooden board, 1 × 4, 4-1/2 ft long

Test Equipment

1	Digital voltmeter (optional)
1	Dual-voltage dc power supply
1	Dual-trace oscilloscope
1	Frequency counter
2	Function generator
1	rf signal generator

Other Equipment

1	AM radio receiver
1	Receiver capable of tuning the 10- or 40-meter shortwave bands
1	TV remote control

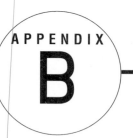

APPENDIX B

SEMICONDUCTOR PINOUTS

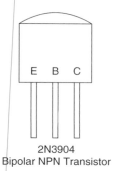

2N3904
Bipolar NPN Transistor

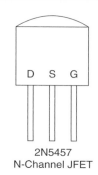

2N5457
N-Channel JFET

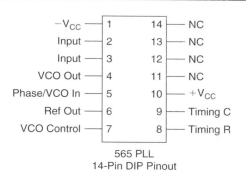

565 PLL
14-Pin DIP Pinout

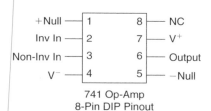

741 Op-Amp
8-Pin DIP Pinout

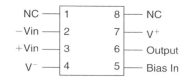

3080 Operational Transconductance Amplifier
8-Pin DIP Pinout

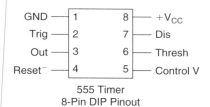

555 Timer
8-Pin DIP Pinout

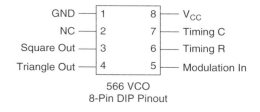

566 VCO
8-Pin DIP Pinout

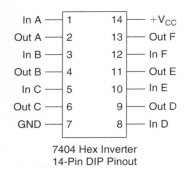

7404 Hex Inverter
14-Pin DIP Pinout

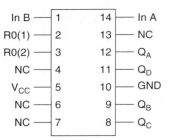

7493 Binary Counter
14-Pin DIP Pinout

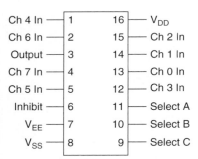

4051 8-Channel Analog Multiplexer
16-Pin DIP Pinout

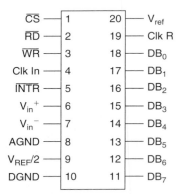

ADC0804 A/D Converter
20-Pin DIP Pinout